Mushrooming with Confidence

Alexander Schwab

Alexander Schwab grew up in Switzerland and was awarded a master's degree in philosophy and history at Aberdeen University. His main interests are all aspects of fishing, hunting, and mushrooming. He lives in the beautiful Emmental region of Switzerland and when not fishing, hunting, or mushrooming he is thinking or writing about them.

Mushrooming with Confidence

A Guide to Collecting
Edible and Tasty Mushrooms

Alexander Schwab

Consultant: Roy Mantle

Skyhorse Publishing

First published by Merlin Unwin Books Ltd. (Ludlow UK) in 2011.

Skyhorse Publishing books may be purchased in bulk at special discounts for sales promotion, corporate gifts, fund-raising, or educational purposes. Special editions can also be created to specifications. For details, contact the Special Sales Department, Skyhorse Publishing, 307 West 36th Street, 11th Floor, New York, NY 10018 or info@skyhorsepublishing.com.

Skyhorse® and Skyhorse Publishing® are registered trademarks of Skyhorse Publishing, Inc.®, a Delaware corporation.

www.skyhorsepublishing.com

13

Library of Congress Cataloging-in-Publication Data is on file.
ISBN: 978-1-62087-195-9

Printed in China

Contents

Mushrooming with Confidence

How to Use This Guide and the Essential Rules

1. Read the entire book *twice* in order to be sure you understand the approach.
2. Study the *Gills, Ridges, Tubes,* and *Spines* sections until you're confident you can correctly distinguish these features; then you're ready for a foray.
3. Double-check the mushrooms you have picked, back at home, by completing the positive identification checklist step by step, which is listed at the end of every species featured in this book. You must be able to check off every single box. If you can, you have the right species. If you can't, you might not have, so don't eat it.
4. Only eat mushrooms that you have clearly identified with all the positive ID marks.
5. If a mushroom smells rotten, it is rotten and if it feels soggy, it is soggy.
6. Never, never eat wild mushrooms raw.

The Mushrooming with Confidence Method

What does "edible" mean?

There are few pleasures as exclusive and satisfying as a feast of delicious wild mushrooms. Likewise, there are few pleasures that can be as fraught with anxiety as mushroom hunting. Nothing curbs the appetite more effectively than a nagging worry that what you eat might make you ill—or kill you. Conventional mushroom identification books promise to enable you to identify hundreds, if not thousands, of edible mushrooms. However, "edible" simply implies "not poisonous" and is no indicator of culinary value. Cardboard is "edible," too.

Mushrooming with Confidence focuses on the very best and most common mushrooms and allows you to identify them safely. In this book, you'll find only the mushrooms of the top league, and, luckily, these are easily and safely identified provided you follow the method presented here. If you carry out all the instructions to the letter, you'll enjoy the most delicious mushrooms, untroubled by fear and doubt.

Learn to leave a mushroom

Some people find it really difficult not to pick everything they spot. It might be

It is tempting to pick every beautiful mushroom you come across, but indiscriminate mushroom hunting is not only dangerous, it makes a nonsense of conservation.

edible, they probably think. They're encouraged in this attitude by encyclopedic mushroom identification books displaying a bewildering multitude of mushrooms. This induces indiscriminate mushroom hunting of the "pick first, ask questions later" variety, which is the wrong approach. Back home, attempts to identify the edible ones with the help of a conventional mushroom identification book invariably fail because there is not enough detail to be truly confident you have the right species.

Even advanced mushroom pickers often face many uncertainties. The same mushroom looks different at each stage of its development and different again when conditions are, for example, very wet or very dry.

Look-alike poisonous species

Conventional mushroom guides always include warnings against look-alike poisonous species. This simply adds to the uncertainty. *Mushrooming with Confidence* asks you not to compare but to positively identify the most valuable mushrooms by their unique and unmistakable features. This approach automatically eliminates dangerous or deadly species, provided you play your part correctly. Basically, this means:

1. Be disciplined enough to leave alone most mushrooms you encounter.
2. Look closely at what you see in front of you, not at what you wish was there.
3. Stick to the rules and check off every box on the identification page in this book.

You wouldn't buy a soggy, worm-infested, half-rotten mushroom in the supermarket. So why pick it in the woods? Quality, as seen in this example of perfect charcoal burners, is what the discerning mushroom hunter is after.

Mushrooming with Confidence aims to make you—not a mycologist—but an expert on the best edible mushrooms. That in turn might provoke a deeper interest in the world of mushrooms. In either case, enjoy!

The tools of the trade

The minimum equipment required consists of a knife and a basket or canvas tote for transportation. Mushrooms want to breathe. If you put them in a plastic bag, they'll suffocate in no time at all, and your beautiful, fresh mushrooms will transform into a soggy mess.

If possible, use a proper mushroom knife with an integrated brush for cleaning. As the first cleaning of the mushroom should be done in the woods, the brush is very useful.

The size of mushrooms

The measurements given in this book are average values. The size of mushrooms can vary disproportionately due to weather and growing conditions: If it is, for example, very dry, the mushrooms will be smaller than average. On the other hand, if conditions are perfect (humidity, ideal substrate), you might come across surprisingly large specimens.

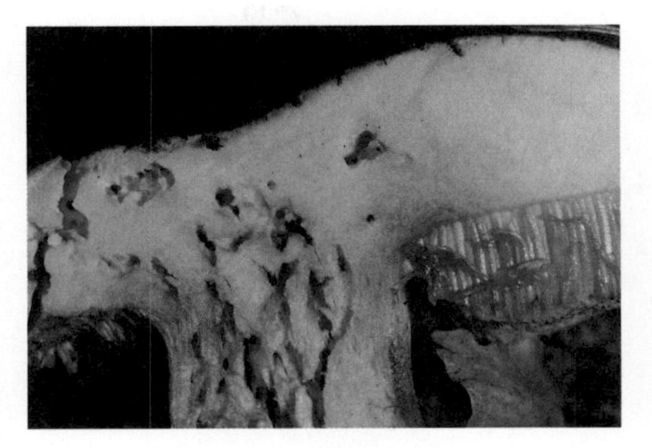

Worm-infested mushrooms

There is no need to discard an otherwise perfect mushroom because of worms. Just cut away the wormy parts. The worm-free pieces are as good as ever—the quality is in no way compromised.

Cutting or pulling?

Cutting mushrooms just above the ground, rather than pulling them up, leaves the mycelium intact. Cutting can actually stimulate the growth of new mushrooms in the mycelium, whereas uprooting can inhibit repeat growth. Sometimes, though, the stem may have features needed for identification, in which case you will have to pull up a specimen.

Raw mushrooms

Some of the mushrooms in this book are only edible after being cooked. Do not eat any wild mushrooms raw as there is a potential risk of catching something nasty from them: Some animals (mice, roe deer) or, worse still, dogs, may have left a mark on them, so avoid busy dog-walking routes.

What Is a Mushroom?

A mushroom is a highly complex organism. Here are the essentials:

Fruitbody

The technical term for what is colloquially called the "mushroom" is the "fruitbody." Mycologically speaking, the real mushroom is the mycelium, which is a fine structure of filaments on or in the forest soil. What we see as mushrooms (the fruitbody) are like the apples on an apple tree.

Cap

The caps of mushrooms have many different forms, but what we're interested in is what is under the cap. Under the cap, you'll find either . . .

Gills, Ridges, Tubes, or Spines

Gills, ridges, tubes, and spines produce, contain, and release the spores. Wind and rain disperse the spores. Given the right conditions, new mycelia and mushrooms form from the spores.

Stem

Mycelium

The mycelium here is visible to the naked eye.

This is a close-up of about four millimeters (a quarter inch) of the same mycelium as seen under a microscope. When conditions are right, the mycelium forms a fruitbody, pushing the mushroom up in order to disperse the spores or to wait for you to come along and pick it.

Gills, Ridges, Tubes, and Spines

Gills, ridges, tubes, and spines provide four neat and useful divisions for identification purposes. Most mushrooms fall into one of these four categories. This is the reason why the first step in the identification process is to determine whether you are dealing with gills, tubes, spines, or ridges. As the following pages show, this is much easier than it sounds.

Gills

Gills are the radiating blades on the underside of the cap. They fan out in a distinctly regular way. Gills have precise forms and come in many colors. Some of them are brittle, some of them are soft. They can be rubbed off or separated from the underside of the cap quite easily. Gills are always attached to the stem or the cap in a uniform way.

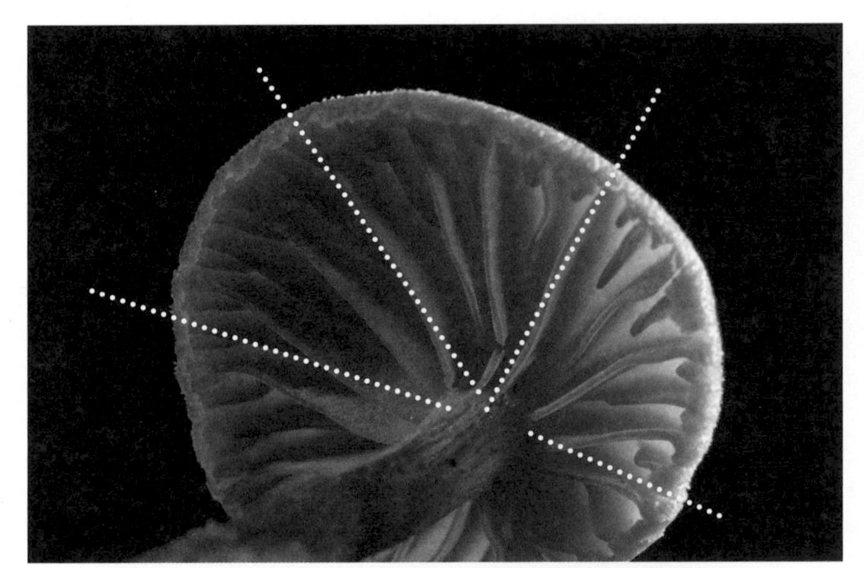

In perfect conditions, the distinctly regular way in which the gills fan out is clearly visible.

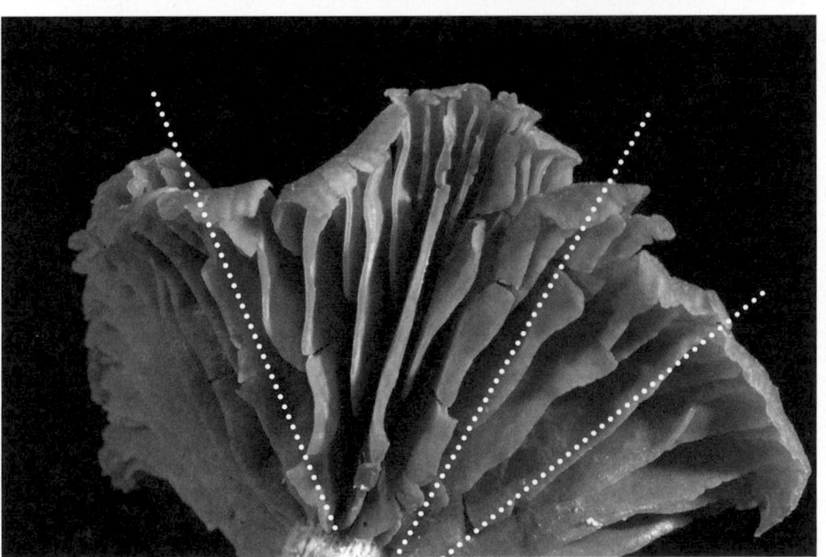

Weather-beaten or old gills might be damaged or broken, but on closer inspection, their regularity will become apparent. Damage or no damage: They run in straight, not wavy, lines.

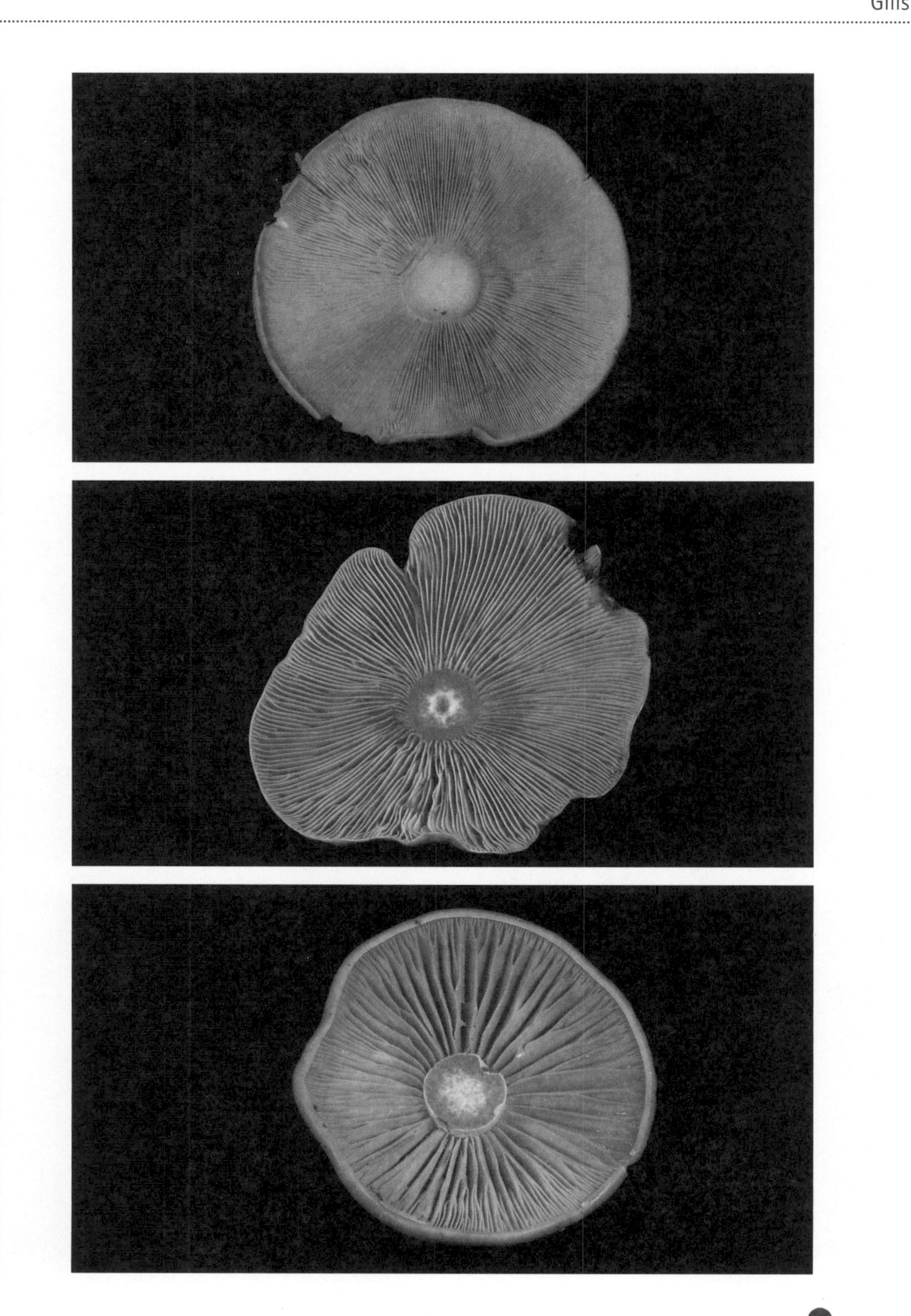

In their regular way, gills differ in spacing and formation. For example:

Gills can fork once or more.

Gills can be mixed: long and short gills.

Gills can be crowded.

Gills can be widely spaced.

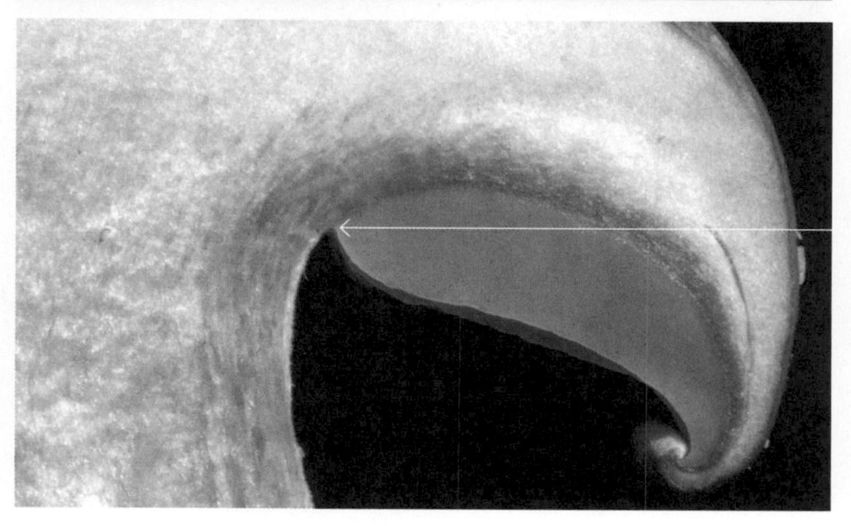

Gills are always uniformly attached to the cap or the stem. They start radiating from the same height along the stem or the same line around the cap.

Ridges

Ridges are on the underside of the cap and form no regular pattern. They have no precise form. They can't easily be rubbed off or separated from the cap or stem.

Ridges are on the underside of the cap. They are not just attached to the cap or stem: They're part of them, which is why you can't rub or pull them off easily.

Ridges are cross-veined, irregular, and don't form a set, uniform pattern. They are not attached to the stem in a distinct and regular way, as gills are: They are part of the stem.

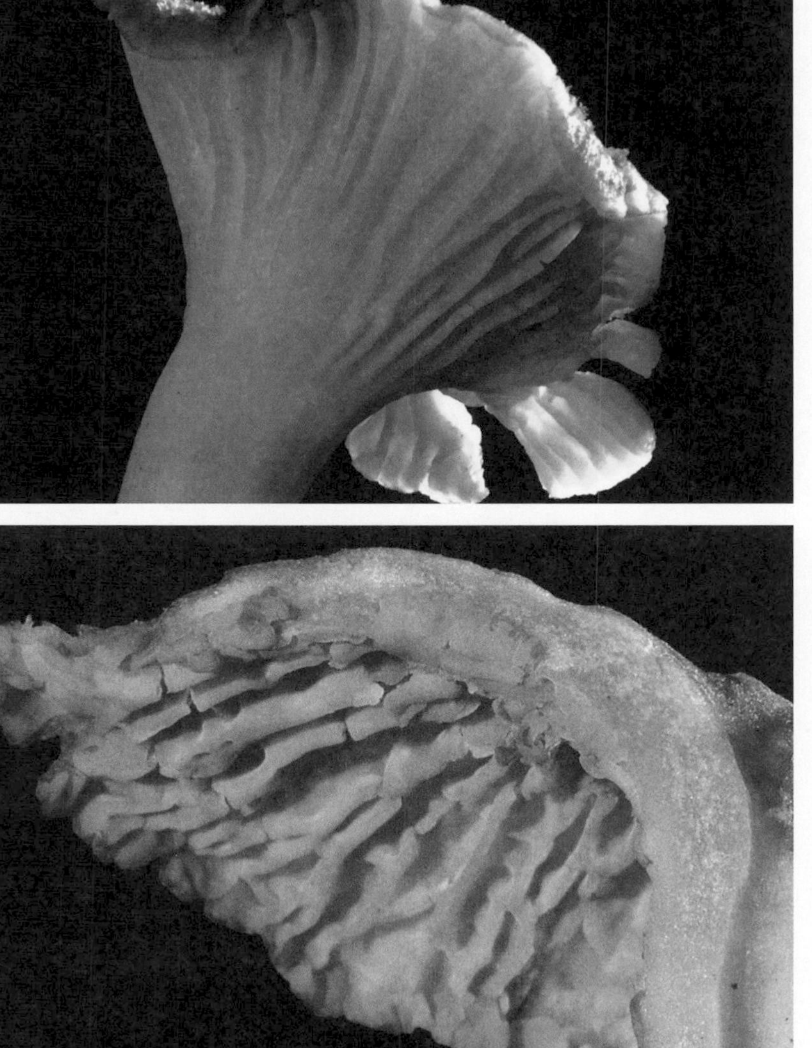

The ridges are part of the stem. There is no distinct pattern in the way they grow out of the stem. Some ridges begin further up, some further down the stem. There is no regularity.

In contrast to gills and tubes, the section of a mushroom with ridges shows no typical features. Because of the irregular nature of ridges, each section will look different, whereas with tubes and gills, you will find the same features each time.

Gills and Ridges in Comparison

Gills

Regular, geometric.

Attached to cap
or stem.

Ridges

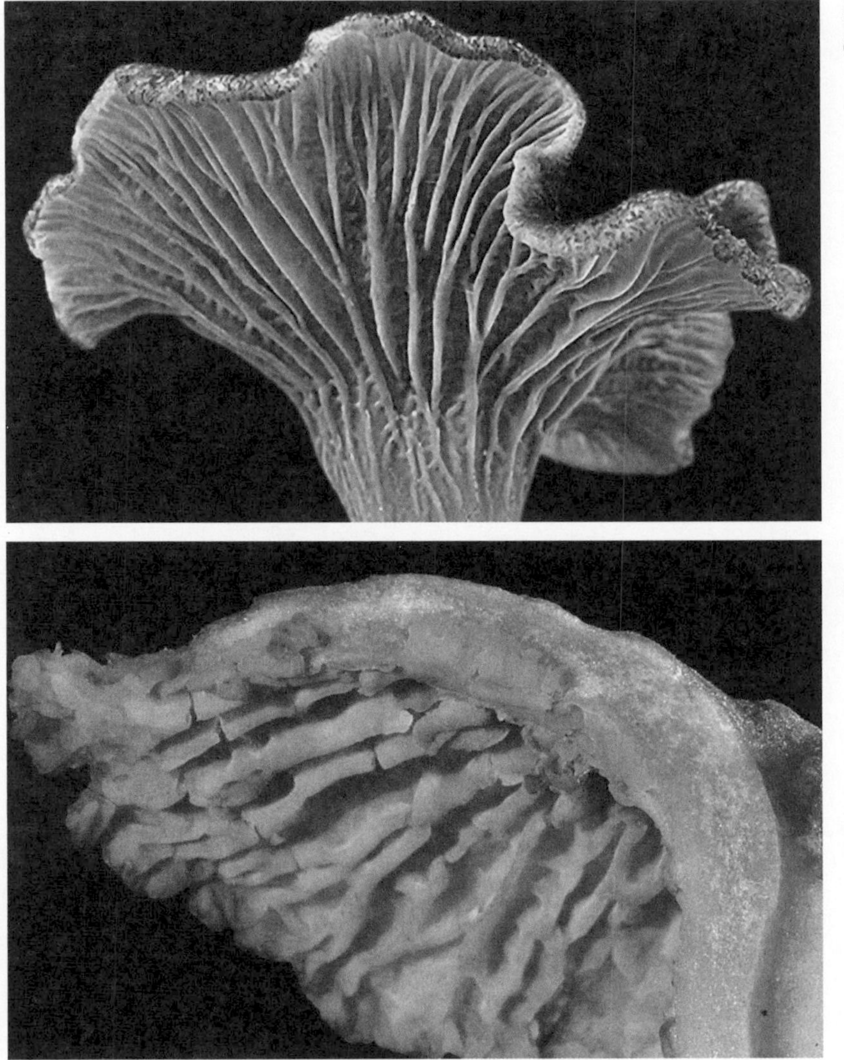

Irregular, cross-veined.

Part of cap and stem.

Tubes

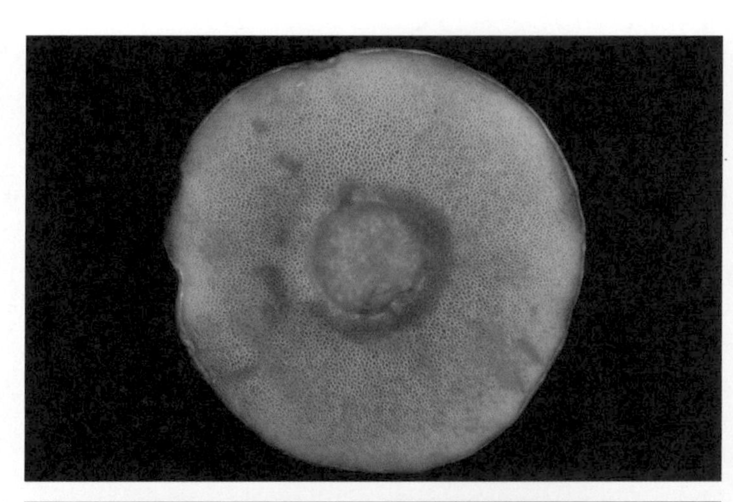

Tubes are fine and tightly packed on the underside of the cap. Without a cross-section, only the lower ends of the tubes are visible as little holes that are called pores. The pores have a sponge-like appearance, which is why the underside of mushrooms with tubes is often referred to as "spongy." Some tubes are more tightly packed than others; in this case, the pores are smaller. Tubes can be removed from the cap easily.

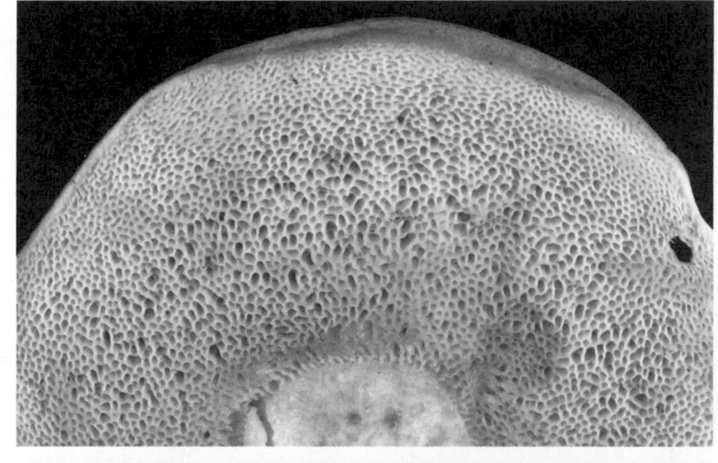

Pores form a regular pattern resembling a sponge.

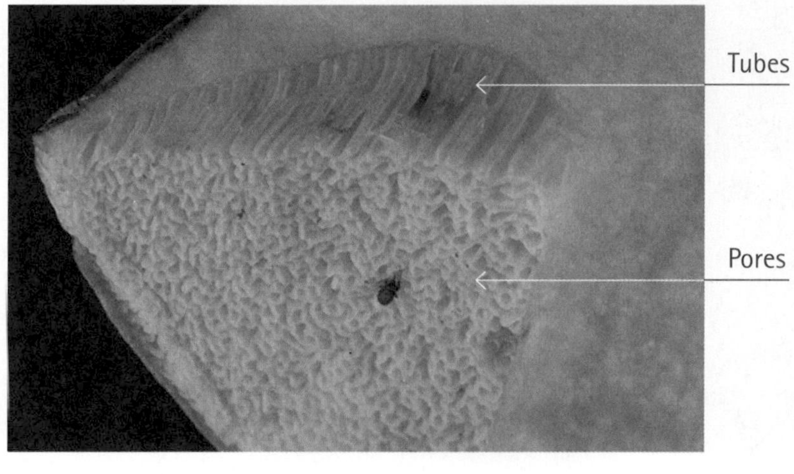

This cross-section shows the tubes and the pores.

Spines

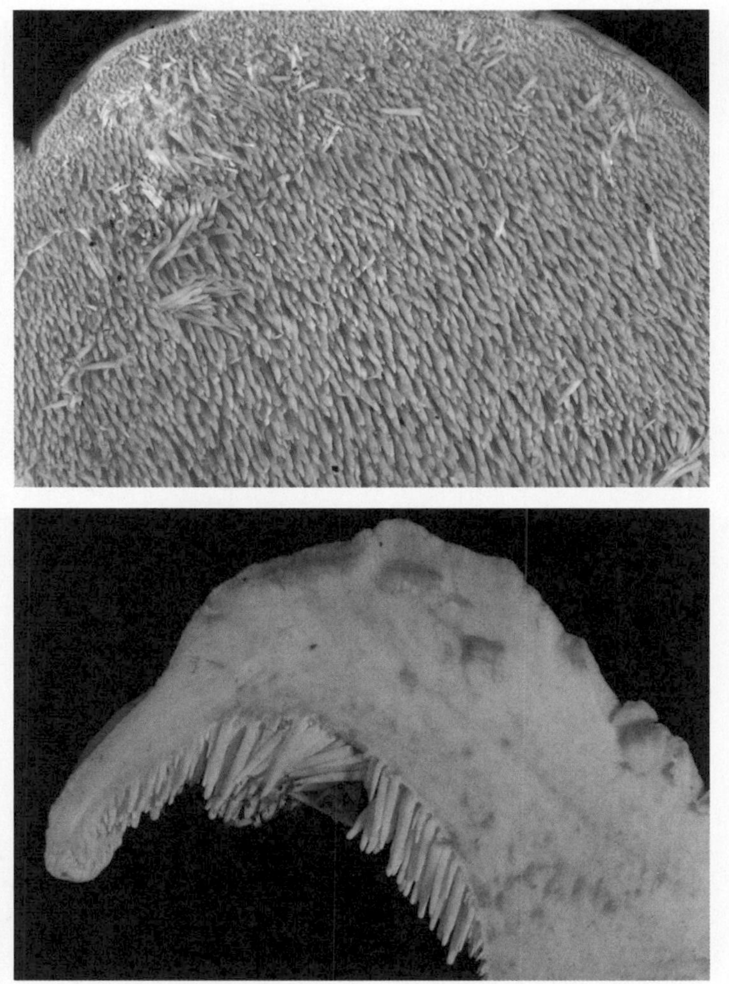

Spines hang like stalactites from the cap.

Positively Identifying Mushrooms

Mushrooms with Gills

The Field Mushroom

Agaricus campestris var. campestris

The field mushroom is probably the most popular mushroom, not least because it is so common. It grows mainly on pastures and grassland but could appear on any grassy, well-fertilized patch of land, even on the lawn. Its culinary value is undisputed, and the flavor of the wild variety is superior to the farmed specimens you can find in the supermarket. If it is a wet and humid summer, start looking for field mushrooms as early as June.

The field mushroom rarely walks alone. It almost always appears in groups or "fairy rings." There might be the odd field mushroom somewhere all on its own—leave it.

The cap of the field mushroom is white. The size of the cap ranges from one to three inches (three to eight centimeters).

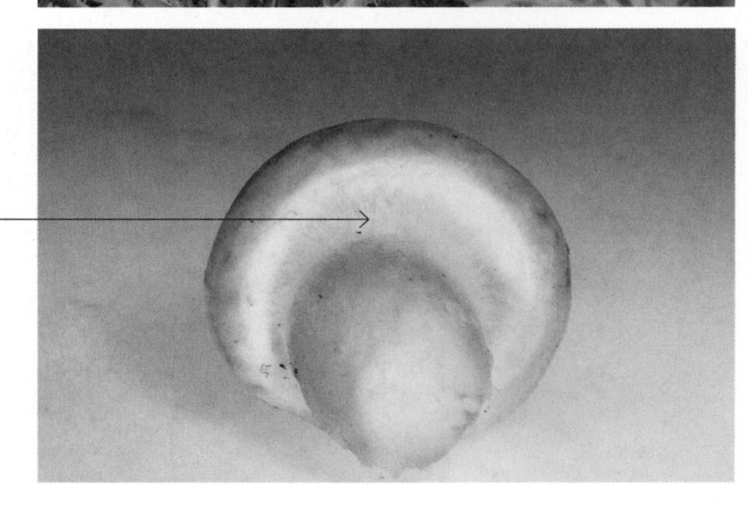

When you find a group or fairy ring of field mushrooms, there will inevitably be specimens at various stages of growth. If the veil is still covering the gills, section the mushroom and ascertain that the gills are pink (see next page).

Veil

The gills of a field mushroom button must not—emphatically not—be white, but light pink, and any bruises on the flesh should be pink. The pink of the bruises varies in intensity.

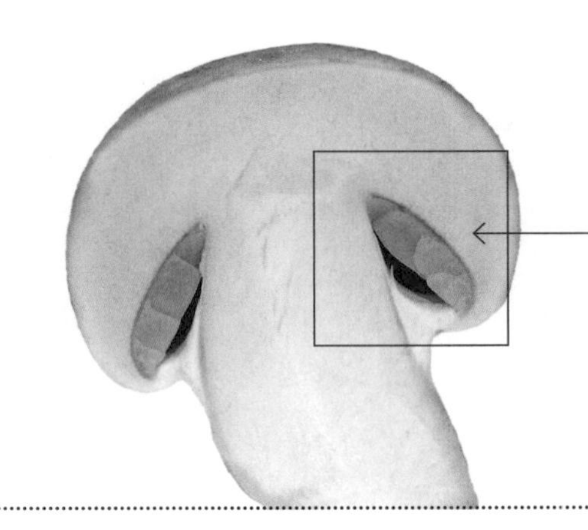

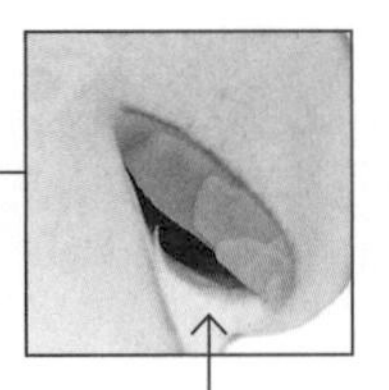

Intact veil covers gills.

Note:
Gills light pink.
Flesh bruises faintly pink.

Veil

As the field mushroom grows, the veil is ripped, leaving a small, fragile ring. This will eventually disappear, but traces of it almost always remain visible on the stem.

Faint traces of veil must be seen on the stem.

The mature field mushroom: deep pink gills and faint traces of the ring left on the stem. The stem is slightly tapered and must not have a bulb at the base. Stems are straight or lightly bent as on this specimen here.

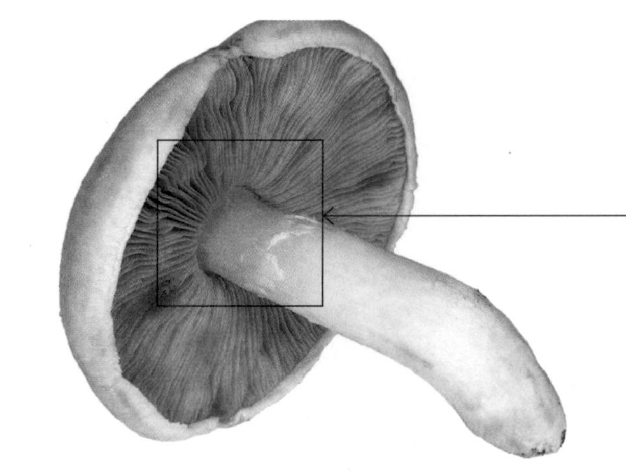

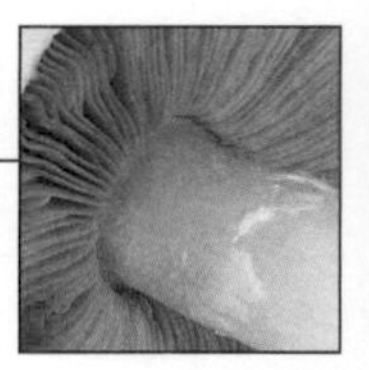

Faint traces of veil remain on stem.

Section of the mature field mushroom: The pink bruising is more intense here and the gills are deep pink. The gills are not attached to the stem.

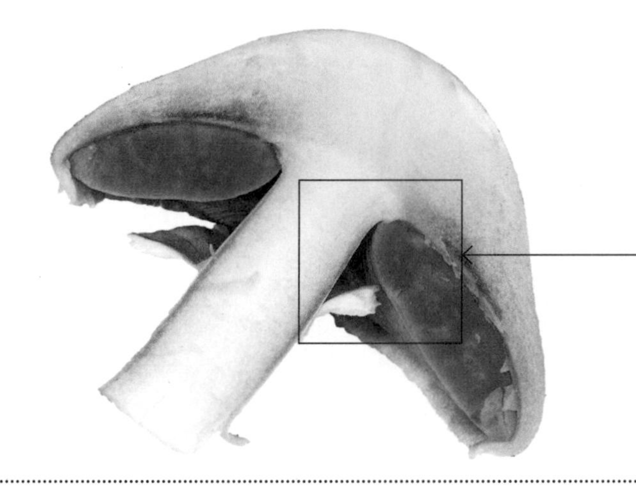

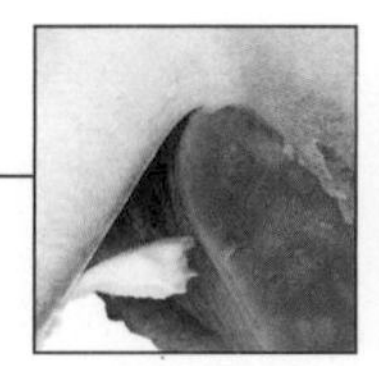

Gills not attached to stem.

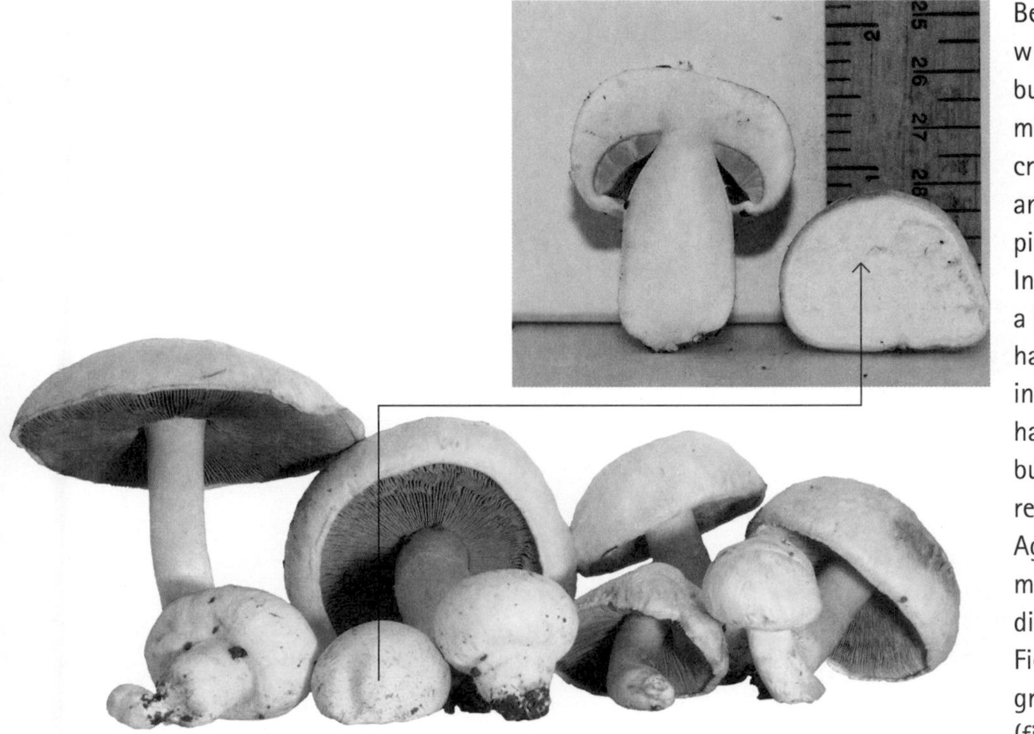

Be extremely careful when collecting button-size field mushrooms. Always cross-section them and make sure the pink gills are visible. In the picture (left), a baby giant puffball has "smuggled" itself into the basket. It is harmless, but some buttons might be really problematic. Again: The gills must be clearly discernible and pink. Field mushrooms can grow up to six inches (fifteen centimeters) in height.

The gills of the field mushroom at different stages: light pink as in the button, and chocolate brown in the late stages of growth. If the gills are dark brown or predominantly black, the field mushroom has lost its culinary value.

Gill color bar

Positive ID Checklist

The Field Mushroom

- ☑ Found in groups or fairy ring
- ☑ Not found in woods
- ☑ White cap
- ☑ Stem tapered, no bulb on base
- ☑ Pink to chocolate-brown gills
- ☑ Gills not attached to stem
- ☑ Flesh turns pink when bruised
- ☑ Gills must match gill color bar (above)
- ☑ On mature specimen: faint trace of ring on the stem

Avg. size across cap:	four inches (ten centimeters)
Season:	June to October
Habitat:	Pastures, meadows, grassland
Tip:	Horse pastures are always very promising
Culinary rating:	10 out of 10

Campestris means "in the plains"

The Wood Blewit

Lepista nuda

If conditions are right, the wood blewit will begin to fruit around the time of the first frost and continue into winter in mild climates. The normal season for this excellent mushroom is September to December. It is quite common and can be found in coniferous and deciduous forests, as well as around compost and brush piles. Apart from its culinary value, it is said to reduce blood pressure. There are reports of very occasional allergic reactions to wood blewits. The first time you eat it, try only a little. Like all other wild mushrooms, it should not be eaten raw.

Wood blewits are found in groups or fairy rings.

When young, the caps are an intense mauve with a pronounced purple tinge.

When older, the colors fade. The older specimens get a tan, so to speak.

Just as the intensity of the cap color changes, the gill color changes from a deep purple to a faint mauvish purple. The gills must be mauvish purple and never brown.

Size of cap about six inches (fifteen centimeters).

Size of cap about four inches (ten centimeters).

Size of cap about two inches (five centimeters).

The convex cap of the young mushroom flattens out as it matures.

The cap of the ideal wood blewit for the kitchen measures between two inches (five centimeters) and four inches (ten centimeters). Wood blewits smell "nice" (e.g. "fruity," "sweet")—they never have a pungent, penetrating odor.

The stem has clean vertical streaks in silvery mauve-purple and is slightly fibrous. There are no brown spots or smears on the stem.

The flesh must be pale mauve-purple—never brown or yellow-brown.

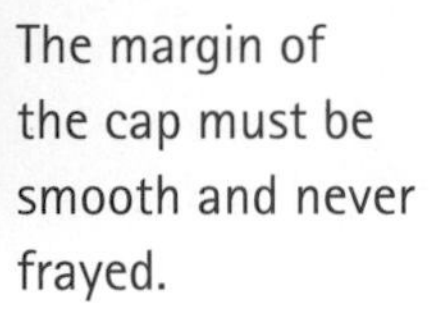

The margin of the cap must be smooth and never frayed.

Cap color bar

Positive ID Checklist

The Wood Blewit

- ☑ Found in the woods or near compost and brush piles
- ☑ Found in groups or fairy ring
- ☑ Cap matches cap color bar (above)
- ☑ Margin of cap smooth, not frayed
- ☑ Stem vertically streaked (silvery mauve-purple)
- ☑ No brown spots or smears on stem
- ☑ Flesh pale mauve-purple
- ☑ Gills match gill color bar (below)
- ☑ Size of cap two to four inches (five to ten centimeters)
- ☑ No pungent odor

Avg. size across cap:	four inches (ten centimeters)
Season:	September, October, November, December
Habitat:	Coniferous, deciduous, and mixed woods
Tip:	Tends to grow in the same places
Culinary rating:	9 out of 10

Nuda means "naked"

Gill color bar

The Shaggy Mane

Coprinus comatus

Luckily, this distinct and delicious mushroom is quite common. Although single specimens are sometimes found standing alone in the middle of nowhere, they usually appear in masses, especially along roadsides and also on disturbed ground and sometimes in the woods. The shaggy mane is also called "the lawyer's wig," a name that aids safe identification of a mushroom which is easy to recognize anyway.

The cylindrical cap covered with curly scales is typical of the shaggy mane.

The shaggy mane grows up to eight inches (twenty centimeters) in height. Do not pick anything you take to be a shaggy mane that is smaller than two inches (five centimeters) in height.

On the left, the perfect specimen for the table. On its right, one that's too far gone.

The stem is hollow. Discard it: It is tough. As long as the gills are white, the shaggy mane is fine to eat.

The quick demise of a beautiful mushroom: The gills liquefy, becoming a black, inky substance. In these stages, the shaggy mane is not fit for consumption.

Young

Mature

Gill color bar

Positive ID Checklist

The Shaggy Mane

- ☑ Found in groups
- ☑ Cylindrical cap
- ☑ Curly scales on cap
- ☑ Hollow stem
- ☑ Minimum height two inches (five centimeters)
- ☑ Gills match gill color bar (above)

Avg. height fit for consumption:	four inches (ten centimeters)
Season:	March to October
Habitat:	Grassland, parks, wayside verges
Tip:	Do not store or transport for a long time. The quality deteriorates by the hour.
Culinary rating:	9 out of 10

Comatus means "shaggy"

The Parasol

Macrolepiota procera

Picking small specimens of the parasol isn't entirely safe because the identifiable features are not fully visible. The specimen on the left is the ideal parasol from the culinary and safety point of view. Parasols show mostly in groups or fairy rings. Where there is one, there is usually another.

These are fine specimens and make an excellent meal.

The general color is brownish cream to white. The paler the mushroom, the more distinctive the snakeskin pattern on the stem.

The cap can measure up to fourteen inches (thirty-five centimeters).

It's all about size: They push up to sixteen inches (forty centimeters).

The gills are off-white to cream and do not change color when touched or bruised. The gills are attached to the cap, not to the stem.

The ring on the stem is white on top and brown underneath and can be slid up and down easily.

The parasol and the shaggy parasol look very similar to the green-spored parasol (*Chlorophyllum molybdites*), which can cause gastrointestinal issues. You can avoid picking this poisonous mushroom by looking for gills that turn green with age and the corresponding green spore print and spore color.

The snakeskin pattern on the stem below the ring is typical of the parasol mushroom.

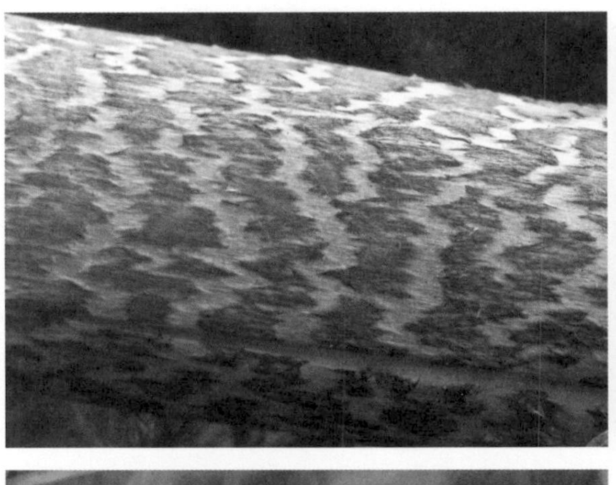

The stem is hollow (right) and forms a mycelium-covered bulb at the base (far right). The bulb is an integral part of the stem. The stem is fibrous and no good for eating. Discard, or dry and crush with a pestle and mortar for a fine mushroom powder for seasoning.

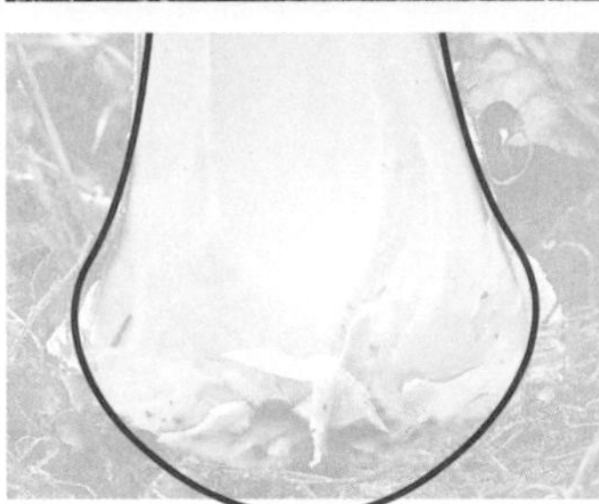

Stem and bulb must be one smooth piece.

Positive ID Checklist

The Parasol

- Over six inches (fifteen centimeters) in height
- Gills do not bruise
- Gills attached to cap, not to stem
- Snakeskin pattern on stem
- Movable ring
- Hollow, fibrous stem
- Bulb integral part of stem
- Gills must match gill color bar (above)

Avg. height ideal for consumption:	eight inches (twenty centimeters)
Season:	July to October
Habitat:	Wood clearings, edges of woods, wayside verges
Tip:	Batter and deep-fry the cap for a special treat
Culinary value:	9 out of 10

Procera means "tall"

The Shaggy Parasol

Chlorophyllum rhacodes

Only collect the shaggy parasol in deciduous or coniferous woods. Some people can get an upset stomach (allergic reaction) from the shaggy parasol, so only eat it in small amounts first. Most people are fine with it, as with all the mushrooms in this book. But a small quantity to start with is always a good idea, as with any new food.

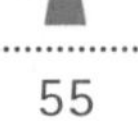

Shaggy cap, hence the name "shaggy parasol." That, however, is not its key identifying feature.

For safe identification, the two-tiered ring should be visible.

The gills are off-white to cream and bruise red. The gills are attached to the cap, not the stem. The flesh of the cap also bruises red.

The stem is hollow and bruises red.

The stem forms a bulb that is covered by mycelium. The bulb is a smooth, integral part of the stem.

The intensity of the red bruising varies and also fades after the initial cutting.

The stalk is whitish, smooth, and shows no particular pattern. The ring is two-tiered and easily movable.

Bruise color bar

Positive ID Checklist

The Shaggy Parasol

- ☑ Minimum height to pick: four inches (ten centimeters)
- ☑ Gills off-white and bruise red
- ☑ Gills attached to cap, not to stem
- ☑ Two-tiered ring on stem movable
- ☑ Hollow stem that bruises red
- ☑ Bulb smooth and part of stem
- ☑ Red bruise matches bruise color bar (above)
- ☑ Found in woods

Avg. size across cap:	four to six inches (ten to fifteen centimeters)
Season:	August to October
Habitat:	Deciduous and coniferous woods
Tip:	Do not pick in gardens or parks
Culinary rating:	9 out of 10

Rhacodes means "ragged" or "tattered"

The Oyster Mushroom

Pleurotus ostreatus

Apart from the fresh air, the reward for braving the frosty winter weather for a woodland walk could be the oyster mushroom. Oyster mushrooms need the frost to fructify (except for some species of oyster mushrooms that fruit earlier in the season. For example, *Pleurotus populinus* is seen on poplar trees in late spring and early summer, and *Pleurotus polmunarius* can be seen in the summer on maple and beech trees). They grow wild on tree trunks, but, since they can easily be farmed commercially, they are often found in supermarkets. The name reflects the fact that the structure of the oyster mushroom resembles oyster banks. Just as fresh oysters are a delicacy, so are fresh oyster mushrooms, and like oysters, they should be consumed as soon as possible after harvesting. Opinions differ as to the oyster mushroom's suitability for freezing or drying. There is no difference of opinion, however, about which of the various species of oyster mushrooms is best: From a culinary angle, the oyster mushroom portrayed here (*Pleurotus ostreatus*) is the only true oyster mushroom.

Tree trunks, stumps, and logs are the places to look for the oyster mushroom. The common beech is probably the best bet, but oyster mushrooms can also be found on other deciduous trees.

Clusters of oyster mushroom viewed from the side, from the front, and from below. The fan-shaped caps overlap in tiers. The gills are white or off-white. The oyster bank structure is the key identifying feature.

The stems are usually short and stubby and often fused together. They are tough: Discard them.

The gills are white to off-white and run down the stem—if there is a distinct stem at all.

The gills turn brownish when old. At this stage, the oyster mushroom is no longer fit for consumption. But there is always next year. . . .

Cap size is usually between two inches (five centimeters) and eight inches (twenty centimeters) but in exceptional cases can be up to sixteen inches (forty centimeters). From the culinary point of view, the smaller cap sizes are preferable because the bigger the cap, the less tasty it is. The cap is smooth, without a veil or traces of a veil (see page 35).

The cap colors of oyster mushrooms vary considerably and look livelier before cutting. The intensity of the colors fades as they dry.

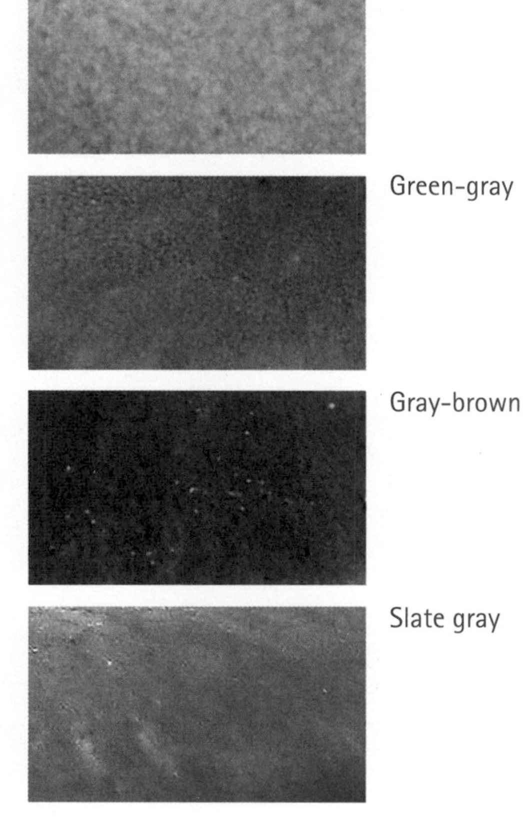

Dove gray-blue

Green-gray

Gray-brown

Slate gray

Gill color bar

Positive ID Checklist

The Oyster Mushroom

- ☑ Oyster bank structure
- ☑ Fan-shaped caps
- ☑ No veil or traces of veil
- ☑ Gills correspond to gill color bar (above)

Avg. size cap:	two to eight inches (five to twenty centimeters)
Season:	November to March
Habitat:	Tree trunks, stumps, logs of deciduous trees
Tip:	Grows year after year in the same place
Culinary rating:	9 out of 10

Ostreatus means "oyster"

The Charcoal Burner

Russula cyanoxantha, Russula variata

The charcoal burner is a very common and excellent mushroom. Apparently slugs also appreciate its mild flavor, which is why tattered specimens seem to be more common than unblemished ones. The charcoal burner belongs to a group of mushrooms called "brittle gills," but luckily it has a unique identification feature: Its gills are not brittle. They are soft.

Brittle gills

You are looking for a mushroom without brittle gills.

First eliminate the look-alike brittle gills. Brittle gills break off when you move a finger over them and apply pressure. The flesh of all brittle gills is brittle and crumbly.

The essential difference between the charcoal burner and other brittle gills is that the charcoal burner's gills are soft and flexible and feel greasy to the touch. They do not break off when pressure is applied; instead, they either resume their original position or stick together.

Note:
With the charcoal burner:
– Gills are white
– Gills are soft
– Gills are greasy to the touch
– Gills do not break under pressure!

The stem of the charcoal burner is smooth and has no ring. It is cylindrical and tapered toward the base. As a rule, the stem is pure white but there is sometimes a light, lilac-purplish flush.

The stem is not fibrous and breaks with a distinct "snap."

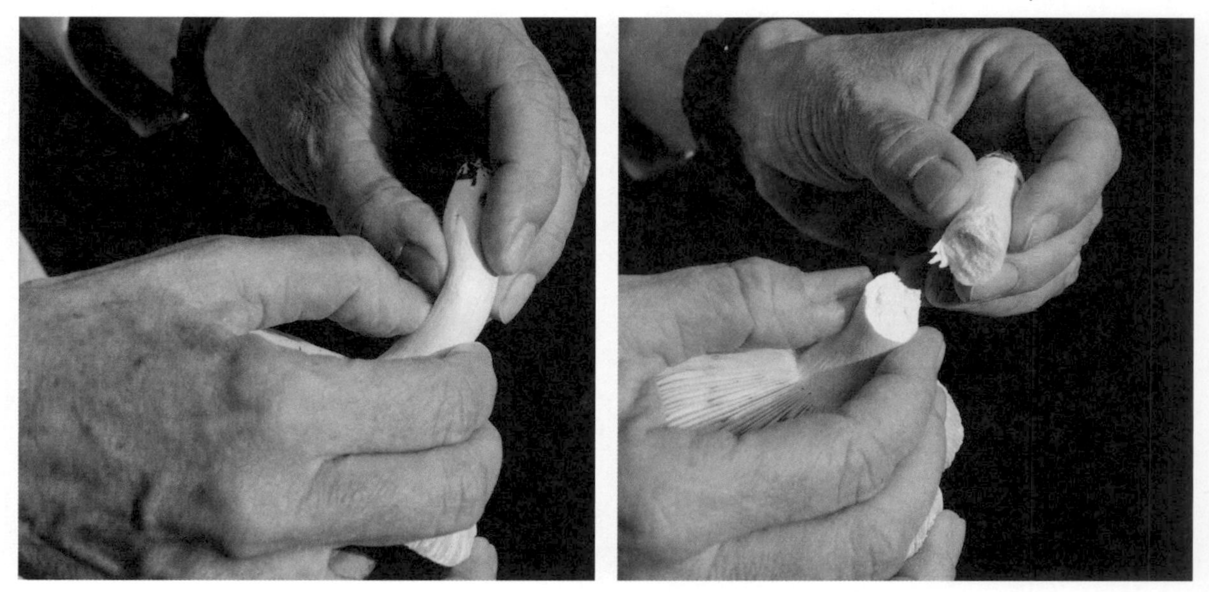

The cap colors vary considerably. There are, however, two main shades: lilac-purple and greenish.

The flesh of the cap is brittle and crumbly.

The skin of the cap can be removed and reveals the lilac-purple-tinged flesh. The stem, too, is sometimes a slightly lilac-purple color.

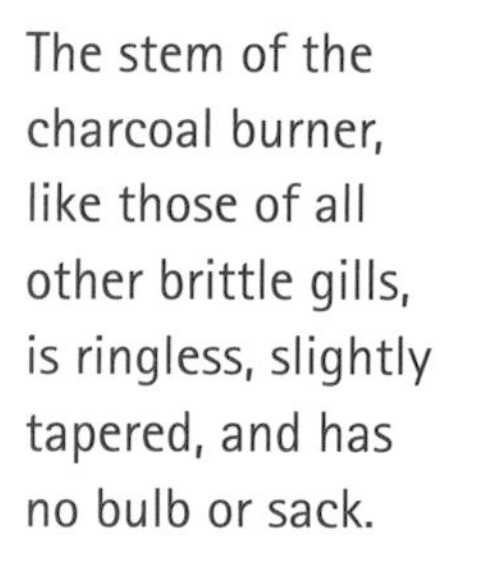

The stem of the charcoal burner, like those of all other brittle gills, is ringless, slightly tapered, and has no bulb or sack.

Safe identification begins at the middle-age stage, fourth from the left, when you can run your finger over the gills. Buttons can't be safely identified.

An exceptionally beautiful and pristine specimen. Note the light flush of lilac-purple on the stem.

Cap color bar
lilac-purple shade

The Charcoal Burner

- ☑ Stem cylindrical and tapered
- ☑ Stem has no bulb or sack
- ☑ Stem is ringless
- ☑ Stem breaks with a "snap"
- ☑ Lilac-purple tinged flesh when cap is peeled
- ☑ White flexible gills
- ☑ Gills do not break under pressure
- ☑ Cap matches one of the cap color bars

Avg. size across cap:	four inches (ten centimeters)
Season:	July to October
Habitat:	Coniferous and deciduous woods but mainly beech
Tip:	Always check on the spot for worms . . .
Culinary rating:	10 out of 10

Cyanoxantha means "blue-yellow"

Cap color bar
greenish shade

The Amethyst Deceiver

Laccaria amethystea

What an attractive little mushroom! Although its taste doesn't fully live up to its aesthetic appeal, it is nevertheless a welcome addition to any mixed mushroom dish. It's purple all over and easy to identify, but as with all other species in this book, proceed systematically and check off every box in the ID checklist.

The purple may vary in intensity depending on the degree of humidity. But the mushroom you pick must be clearly and fully purple—despite the whitish fibrous streaks in the stem.

The amethyst deceiver is often found with the trumpet chanterelle (see page 139).

Cap size across: ½–2½ inches (1–6 cm).

It is equally at home in coniferous and deciduous woods.

The arrangement of the gills must look like this.

The stem is fibrous and should be discarded after the completed identification.

After picking, colors fade. Again: Only pick fully purple specimens.

Positive ID Checklist

The Amethyst Deceiver

- ☑ Fully purple cap
- ☑ Fully purple gills
- ☑ Fully purple stem
- ☑ Fibrous stem with whitish streaks
- ☑ Gill arrangement as in picture, top of page 77
- ☑ Gills match gill color bar (above)

Avg. size across cap:	three-quarter to two inches (two to five centimeters)
Season:	August to October
Habitat:	Mainly in deciduous but also in coniferous woods
Tip:	Likes mossy patches
Culinary rating:	6 out of 10

Amethystina means "amethyst" or "purple"

Mushrooms with Tubes

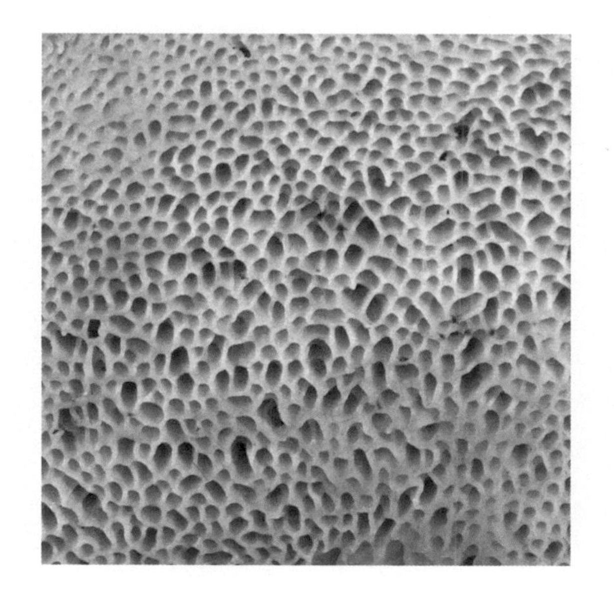

The King Bolete

Boletus edulis

The king bolete is the king of mushrooms. Some truffles are more rare, more expensive, and confined to relatively small areas of the world. The king bolete, on the other hand, reigns supreme everywhere. King boletes can be found in the same location year after year. They can disappear in certain areas for a couple of years only to return spectacularly in masses. Some people gather only king boletes, which is understandable because of their beauty and taste, and the thrill of finding them on the same spot time after time. King boletes vary greatly in general appearance, color, and size. The size of the little fellow in the foreground of the picture above is about two inches (five centimeters) in height whereas the giant is about fourteen inches (thirty-five centimeters) in height. If conditions are right (starting off with a wet and humid spring or early summer), you should begin to look for king boletes as early as the beginning of July. They can continue to the end of October.

These are summer king boletes. All king boletes, however, whether summer or autumn, show a fine white mesh pattern on the top of the stem right underneath the tubes.

A classic autumn king bolete, the king of mush-rooms.

This is the key ID mark. The white mesh pattern must be visible! All king boletes, regardless of their season or stage of development, have a fine white mesh pattern at the top of their stem.

This is what a king bolete ideally looks like when cut. Not a single worm has even looked at this beauty. Neither the tubes nor the flesh change color when cut or touched.

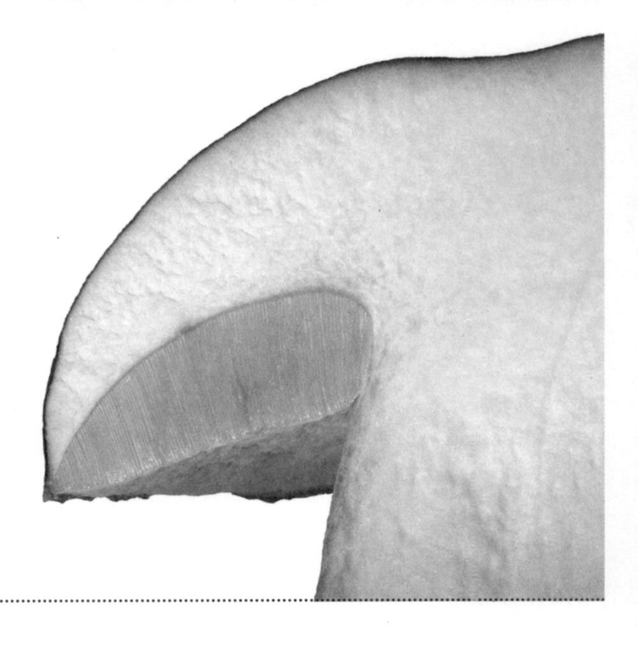

King bolete tubes are off-white and firm when young. Later, they turn yellow-olive and are less firm. In maturity, the tubes are olive. Tubes of mature king boletes are soft as a sponge and the tubes of old ones are soggy and often like those of the big fellow on page 81. This does not always mean that the inside of old king boletes is rotten but, more often than not, this is the case.

King bolete stalks vary greatly in appearance. This specimen clearly shows the white mesh pattern all over the stalk.

It is the top section of the stalk just below the cap that matters:
A mesh pattern must be clearly visible.

Cap color bar

Positive ID Checklist

The King Bolete

- ☑ Tubes, pores, and flesh do not change color when cut or bruised
- ☑ Tubes are off-white, cream, yellow-olive, or olive
- ☑ Pores do not show any pink tinge
- ☑ White network on top of stem
- ☑ Cap matches cap color bar (above)

Avg. size across cap: five inches (thirteen centimeters)
Season: July to October
Habitat: Underneath beech, oak, birch, and pine trees
Tip: This mushroom has a lot of stem! It tastes as good as the cap
Culinary rating: 10 out of 10

Edulis means "edible"

The Red Cracked Bolete

Boletus chrysenteron

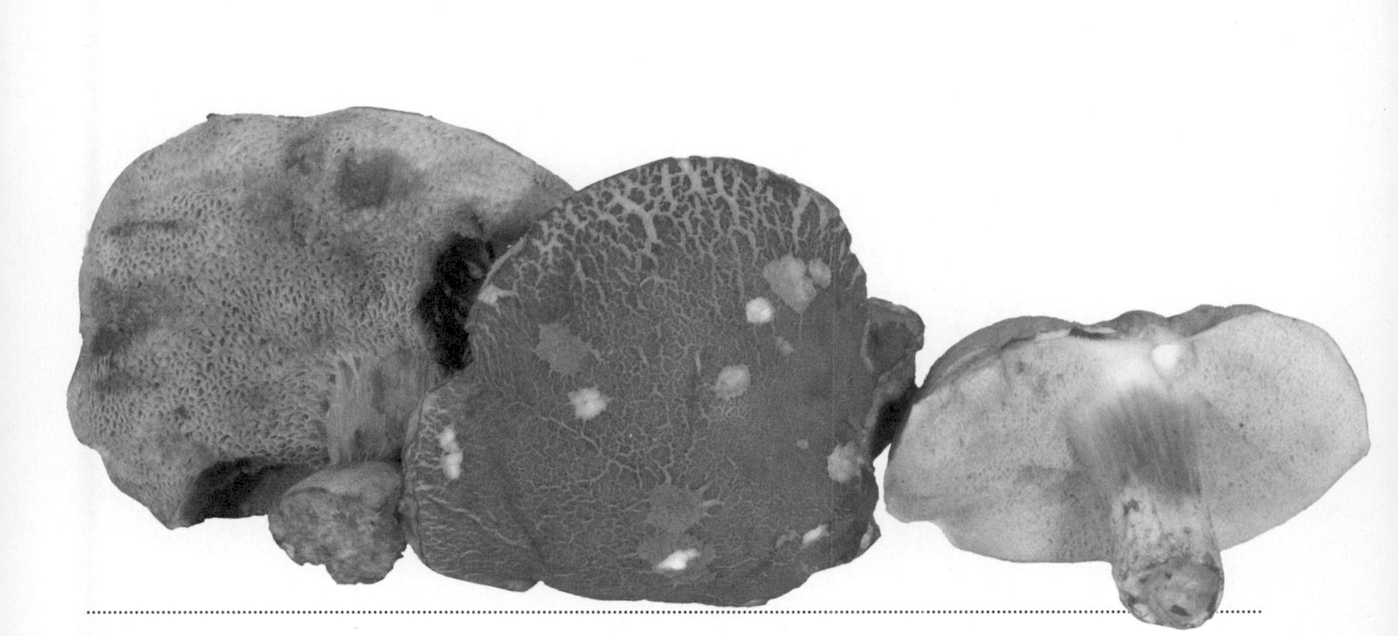

The red cracked bolete is a common and easily distinguished mushroom with tubes. Its name describes its key identification feature perfectly. As the mushroom matures, the cracks become more prominent. Opinions on the culinary value differ significantly. The reason is its distinct fruity smell and taste, which divides people. In any case, the red cracked bolete should only be picked as long as the pores are bright yellow and firm. The ideal size of the cap is about three-quarters of an inch to an inch (two to three centimeters) across. Anything larger will make your dish all slimy.

The color of the cap ranges from a light brown to a velvety dark brown.

Stems vary in color. They can be yellow, yellow with a little red, or yellow flushed with red.

The stem and pores bruise blue. The intensity of the blueing varies. (See picture on the next page.)

This specimen is too old for the kitchen. The pores have lost their brightness. The dull yellow signals that it's no good anymore. When the pores are this color, the mushroom feels soft to soggy.

This is the perfect specimen. It has bright, firm golden-yellow pores and a nice firm cap about one inch (three centimeters) across.

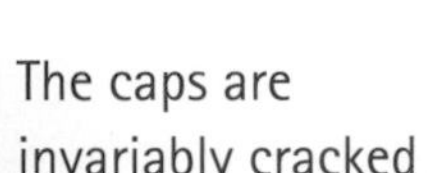

The caps are invariably cracked or slightly damaged. The typical identification mark of the red cracked bolete is the red showing through the cracks or the eaten-away patches.

When cut, the white-to-bright-yellow flesh turns blue. This can be very light blue and confined to certain patches.

Cap color bar

Positive ID Checklist

The Red Cracked Bolete

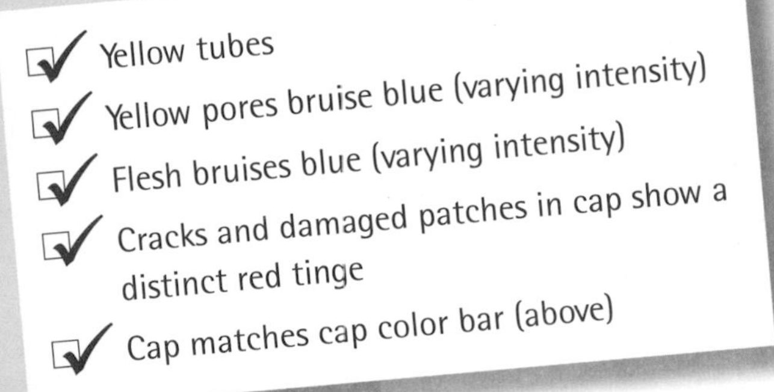

- ☑ Yellow tubes
- ☑ Yellow pores bruise blue (varying intensity)
- ☑ Flesh bruises blue (varying intensity)
- ☑ Cracks and damaged patches in cap show a distinct red tinge
- ☑ Cap matches cap color bar (above)

Avg. size across cap:	two inches (five centimeters)
Season:	June to November
Habitat:	Seems to feel at home everywhere
Tip:	The best are the small specimens (i.e., cap no wider than one inch [three centimeters] across)
Culinary rating:	Anything from 0 to 10 out of 10

Chrysenteron means "with golden-yellow flesh"

The Larch Bolete

Suillus grevillei

It is as if pure gold has grown out of the ground—and where there is one nugget, there are others. It's very rare to find a single larch bolete. More often than not, larch boletes form a fairy ring. Always found near larch trees, this beautiful and delicious mushroom is covered with a yellow veil when young (above left). As the maturing mushroom grows, the veil breaks, leaving a transient ring on the stalk (above, center, and right). Always peel the cap of the larch bolete on the spot because otherwise it will make your basket, and later your cooking, all slimy. In wet conditions, the cap of the larch bolete is always slimy.

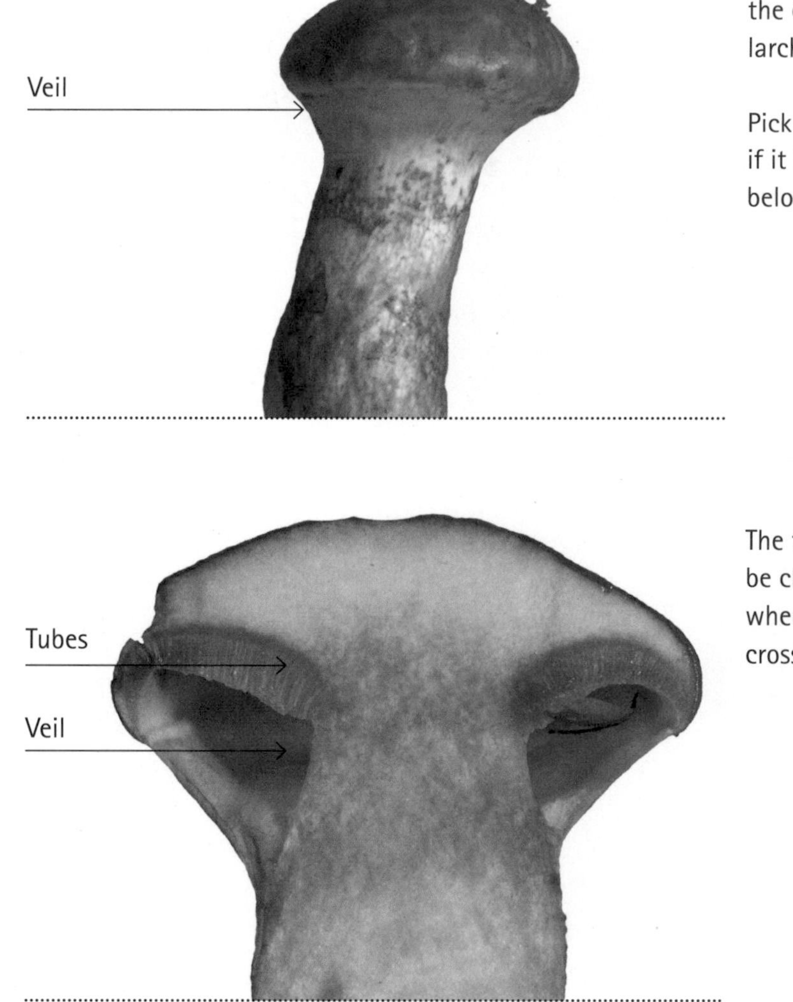

The veil still covers the cap in a "baby" larch bolete.

Pick it and check if it has tubes (see below).

The tubes must be clearly visible when you view a cross-section.

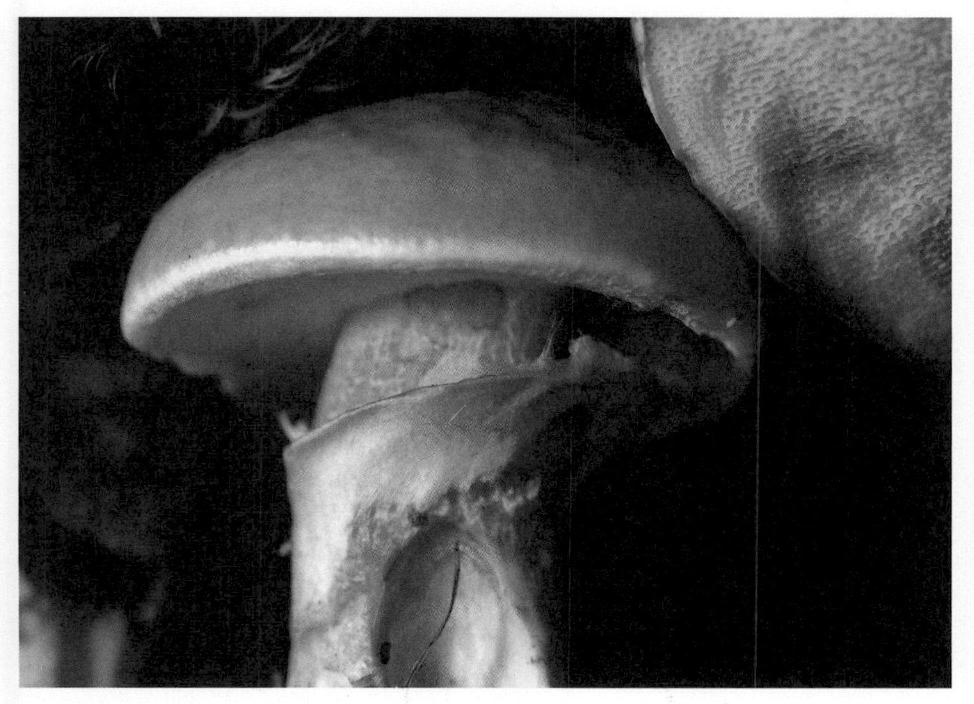

As the mushroom grows, the veil breaks. This picture (left) shows a larch bolete shortly after the breaking of the veil. The ring is transient (i.e., it often drops off), but even in the mature larch bolete you can always see a mark where the ring has been.

Above the ring marks: net pattern (can be very faint).

Below the ring marks: Mature specimens show distinct reddish, rusty streaks on the yellow stem.

If a larch tree is nearby and you see several gold nuggets, they are bound to be larch boletes. The beautifully golden (and slimy when wet) caps are the key sign. Positive identification, however, requires checking off all of the identification marks.

Cap color bar

Positive ID Checklist

The Larch Bolete

- ☑ Found in group or fairy ring
- ☑ Found near larch trees
- ☑ Yellow tubes
- ☑ Solid stem
- ☑ Bright orange-yellow or golden-yellow sticky cap (slimy when wet)
- ☑ Flesh in cap flushes lilac (faint)
- ☑ Young: tubes covered with a veil
- ☑ Intermediate stage: ring visible, faint net pattern above ring zone
- ☑ Mature: ring zone still visible, faint network above ring zone
- ☑ Cap color matches cap color bar (above)

Avg. size across cap:	two and three-quarters inches (seven centimeters)
Season:	June to November
Habitat:	Always near larch
Tip:	Always remove the cap skin
Culinary rating:	9 out of 10

Grevillei refers to the Scottish mycologist R.K. Greville

The Slippery Jack

Suillus luteus

The slippery jack is most often found under the Scots pine or red pine, but can also grow under white pine and other conifers. When wet, the cap of the slippery jack is slimy and off-putting to some people. The cap skin is easily removed, however, and what you get is described by some experts as a “prized” mushroom, while others classify it as merely ”edible.” The truth is probably somewhere in the middle.

Although you might find a lone slippery jack, they are usually found in groups near Scots pine or other conifers.

The cap color is generally chestnut brown (left). The cap can take on a purplish gray hue (right).

Cap size is from one and a half to six inches (four to fifteen centimeters). Always remove cap skin before use—you don't want slime.

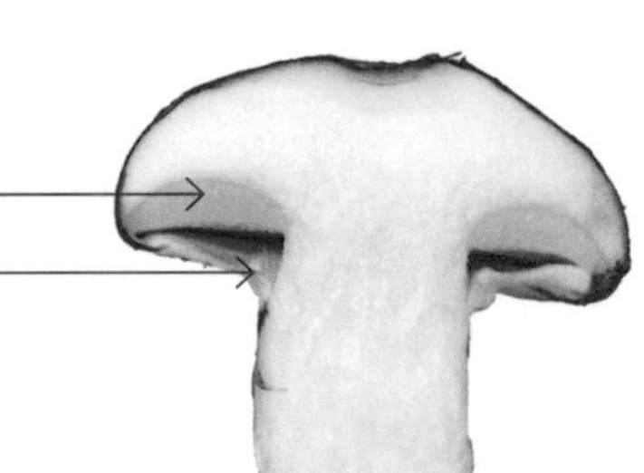

The flesh is white to yellowish-white, the tubes buttery-yellow

Initially, the tubes are covered by a veil that, as the mushroom grows, breaks and leaves a floppy ring.

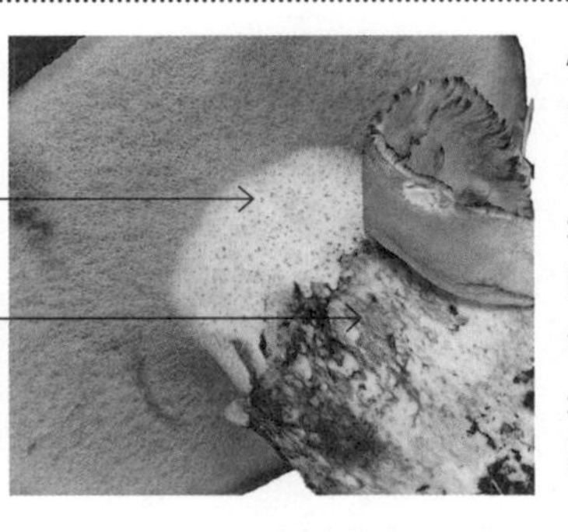

Above the floppy ring, the stem is pale yellow sprinkled with little brown dots. Below the floppy ring, the stem turns purplish brown with age.

Positive ID Checklist

The Slippery Jack

- ✓ Found in groups or fairy ring
- ✓ Buttery-yellow tubes
- ✓ Floppy ring
- ✓ Brown dots above floppy ring
- ✓ Cap matches cap color bar (above)

Avg. size across cap:	three inches (eight centimeters)
Season:	September to October
Habitat:	Coniferous woods
Tip:	Always remove cap skin
Culinary rating:	7.5 out of 10

Luteus means "saffron" or "yellow"

The Bay Bolete

Boletus badius

The bay bolete is an excellent mushroom and luckily is also very common. The main bay bolete months are September and October, but it can sometimes be found as early as June. If you find one, there are bound to be others nearby.

Yellow pores: This is the stage when the bay bolete is at its best. If the pores turn dirty yellow or green (see mushroom on far right, page 103), the mushroom is past its culinary best.

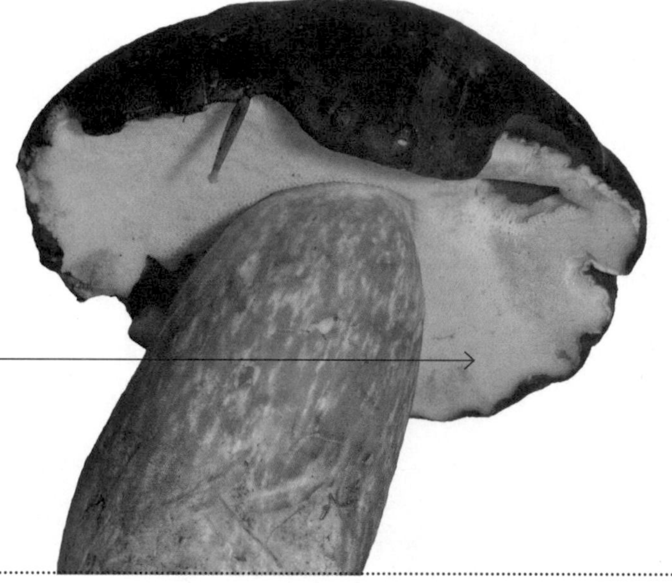

The yellow pores bruise green-blue.

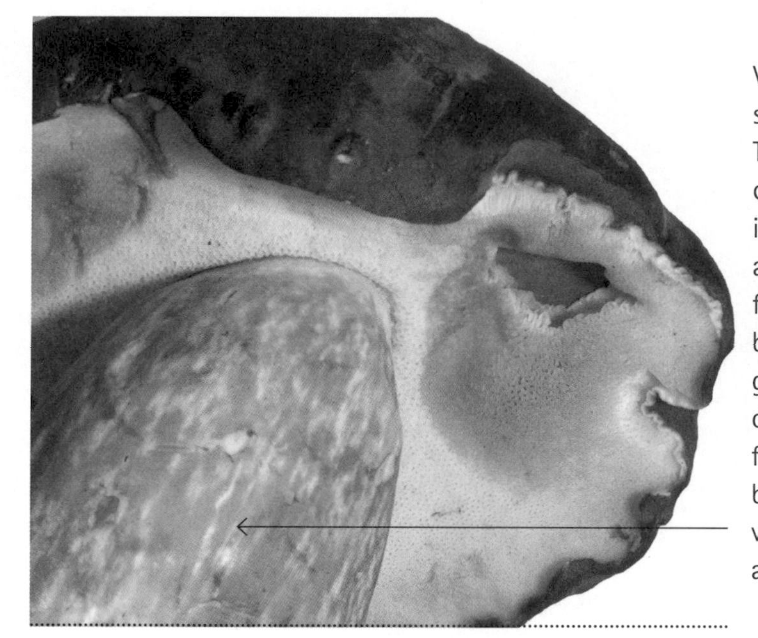

Vertically streaked stem: The background color of the stalk is yellow-brown and is vertically frosted with brown streaks. The general appearance of the stalk ranges from light to dark brown but the vertical streaks will always be visible.

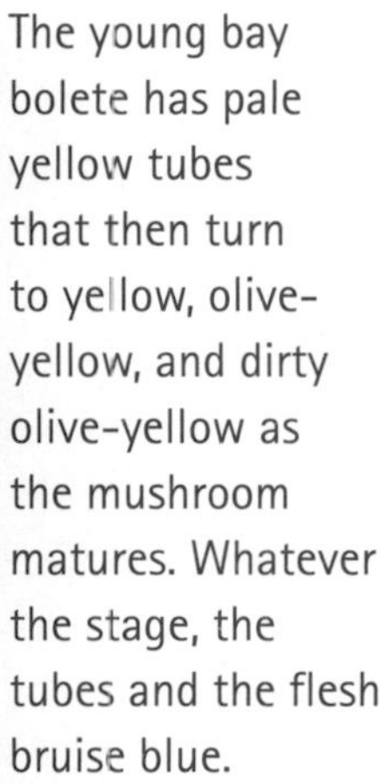

The young bay bolete has pale yellow tubes that then turn to yellow, olive-yellow, and dirty olive-yellow as the mushroom matures. Whatever the stage, the tubes and the flesh bruise blue.

The change in color of the pores to green-blue when bruised varies in intensity. Here it's quite extreme.

The intensity of the coloring varies, but pores, tubes, and flesh will always bruise green-blue.

This is what the ideal bay bolete looks like: yellow pores and tubes, white to white-yellow flesh, without a trace of a worm.

Not as good, but if you discard the soft pores and cut away the wormy parts, it will still be excellent.

The color of the cap will always be "bay" (chestnut brown), but bay has a considerable range.

When wet, the cap will be slightly slimy and this will intensify its color.

When dry, the color of the cap will be duller.

Cap color bar

Positive ID Checklist

The Bay Bolete

- Yellow tubes
- Yellow pores bruise green-blue
- Tubes and flesh bruise blue
- Stem is vertically frosted with brown streaks
- There is no mesh pattern of any description on the stem
- Cap color matches cap color bar (above)

Avg. size across cap:	three and a half inches (nine centimeters)
Season:	June to November
Habitat:	Prefers coniferous woods
Tip:	Remove any soft tubes
Culinary rating:	10 out of 10

Badius means "beautifully brown"

The Birch Bolete

Leccinum scabrum

The birch bolete and its cousin, the orange birch bolete, are the most common of the "rough stalks." Rough stalks are mushrooms with a scaly stem. Birch woods or groups of birch trees are the places to look for them. The birch bolete and the orange birch bolete have numerous relatives, all of which have a black or brown scaly stalk. They are all edible, but it is recommended to stick to the two species presented here. It's a good idea to try eating a small quantity first to check if you are allergic to this mushroom, as some people have reported allergic reactions.

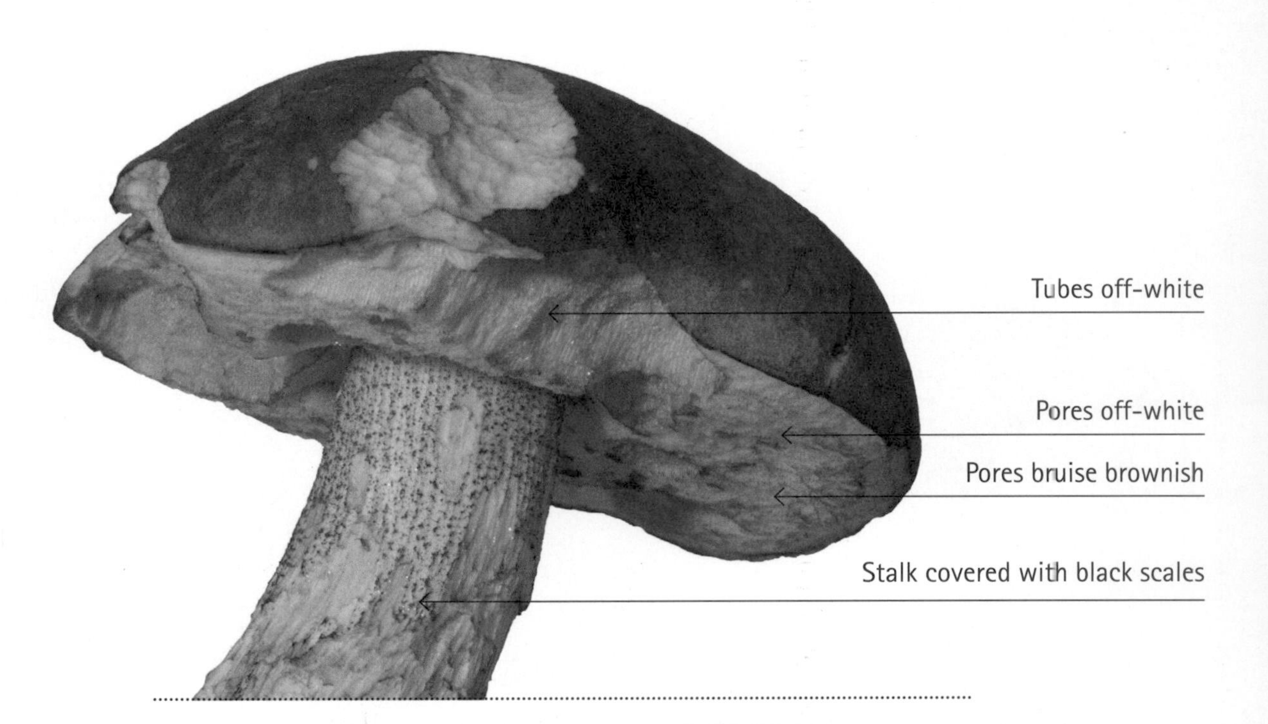

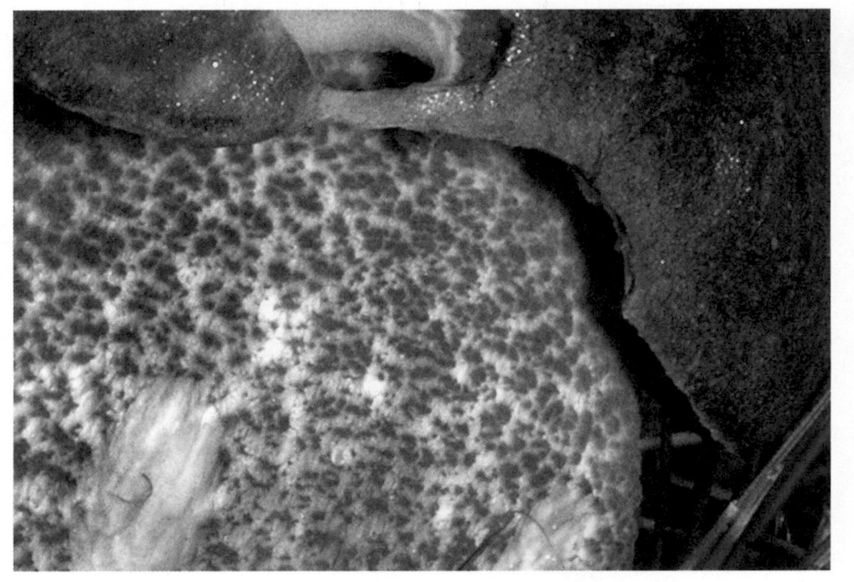

Close-up of stalk, showing black scales.

Pores and tubes change from off-white to a gray-white color as the mushroom matures. As a rule, you should only pick the birch bolete if it is firm and the pores are off-white.

Larger specimens are occasionally still nice and firm, especially when conditions have been very dry.

These are the perfect birch boletes: a feast for the eyes and delicious to eat.

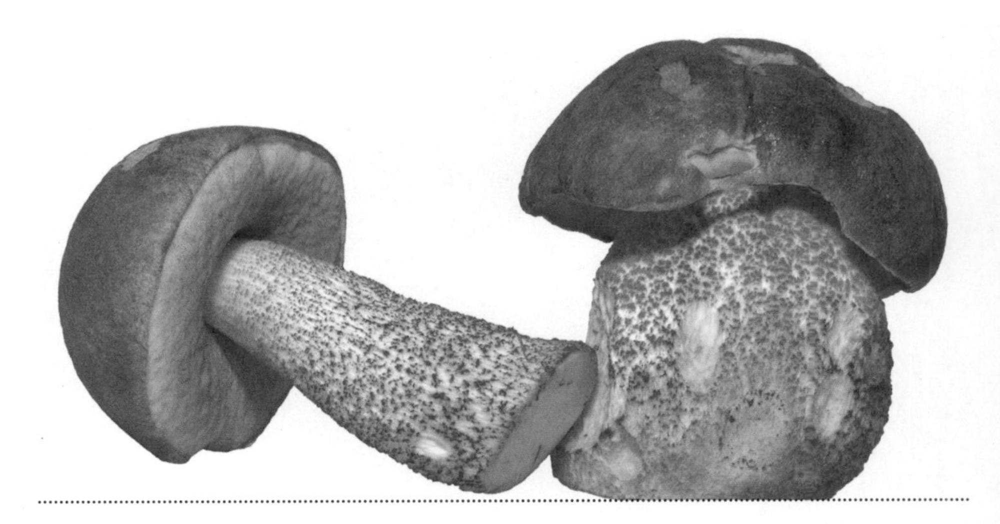

The orange birch bolete, *Leccinum versipelle*, is a beautiful mushroom. Its key identification mark is the stem covered with black scales, and of course the orange cap. When cut, the flesh turns faintly blue at the stalk base, displays traces of wine-red, and finally turns and stays gray. This color combination doesn't look especially appetizing, but be assured: The orange birch bolete is delicious.

The orange birch bolete differs widely in appearance, but the rough, scaly stalk, which is the key identification feature, remains the same.

Positive ID Checklist

The Birch Bolete

- Pores off-white
- Pores bruise brownish
- Stalk with small black scales
- Cap color matches cap color bar (above)

Avg. size across cap:	three and a half inches (nine centimeters)
Season:	July to November
Habitat:	Birch woods, groups of birch, individual birch
Tip:	Discard stem and tubes of larger specimens
Culinary rating:	9 out of 10

Scabrum means "rough"

Mushrooms with Ridges

The Chanterelle

Cantharellus cibarius

The chanterelle is a very popular mushroom, not surprisingly. It is delicious and very common. Since it is imported throughout the year, the supermarket is a good place to familiarize yourself with the chanterelle. There is, however, nothing like the chanterelle you find yourself. They are found in groups or scattered, never in dense clusters. Refrain from picking tiny chanterelles, that is, little yellow buttons that hardly show the identification features. At some point, you'll come across one lonely chanterelle standing in the middle of nowhere for no particular reason at all. Leave it.

The classic color of the chanterelle is egg-yolk yellow, which, however, can vary from light yellow or yellow-orange to yellow-ochre. The chanterelle does not change color when bruised. The entire outside of the mushroom is the same color.

Young chanterelles have a flat cap with a slightly down-curled edge. Do not pick any chanterelles smaller than one inch (three centimeters) high.

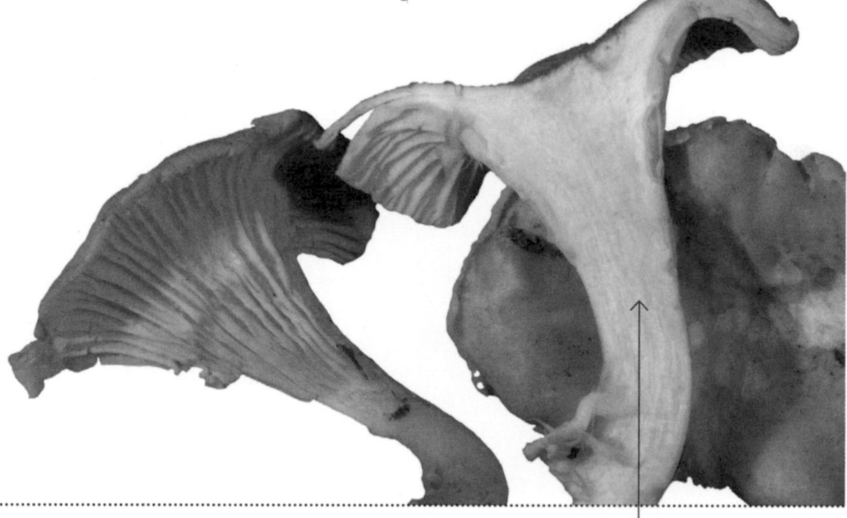

The more mature the chanterelle, the more funnel-shaped it becomes.

The stalk is solid. The whitish to yellowish flesh does not change color when bruised or cut.

Whether young
or mature, the
chanterelle's ridges
must be clearly
visible.

Some mature
forms can look
highly irregular,
but, on closer
inspection,
the funnel shape
will still be there.

Cap color bar

Positive ID Checklist

The Chanterelle

- Young specimen: flat cap, slightly down-curled edge
- Mature specimen: funnel-shaped
- Solid stem
- Found in groups
- Ridges must be clearly visible
- Color matches cap color bar (above)

Avg. size across cap:	two inches (five centimeters)
Season:	July to October
Habitat:	In woods with beech, oak, pine, and birch
Tip:	Watch for mossy patches with little plant cover
Culinary rating:	10 out of 10

Cibarius means "belonging to food"

The Trumpet Chanterelle

Cantharellus tubaeformis

In comparison to the attractively colored chanterelle, the trumpet chanterelle (also known as the autumn chanterelle, or winter chanterelle) looks modest. On closer inspection, however, it is just as beautiful as the chanterelle. The trumpet chanterelle pops up over-night so quickly that you could watch it grow. It is an endearing mushroom, not least because it tends to appear in great numbers and is frost-resistant.

A cluster of trumpet chanterelles.

The cap is brownish and has a depression (small specimen) or a hole connecting to the hollow stem in the center.

Color and shape vary according to age and weather conditions. Each individual trumpet chanterelle has more than one color.

The ridges and the stalk range from brown-yellow to gray-yellow to gray-lilac.

The jagged edges here are due to frost.

The trumpet chanterelle has a hollow stem.

Systematically check each specimen. Can you spot the odd one out here?

Cap color bar

Positive ID Checklist

The Trumpet Chanterelle

- ☑ Found in groups
- ☑ Brownish cap with depression or hole in center
- ☑ Hollow stem
- ☑ Ridges
- ☑ Cap color matches cap color bar (above)

Avg. size across cap:	three quarters of an inch to an inch (two to two and a half centimeters)
Season:	September to November
Habitat:	Deciduous and coniferous woods
Tip:	Mossy banks are a favorite habitat
Culinary rating:	9 out 10

Tubaeformis means "trumpet-shaped"

Mushrooms with Spines

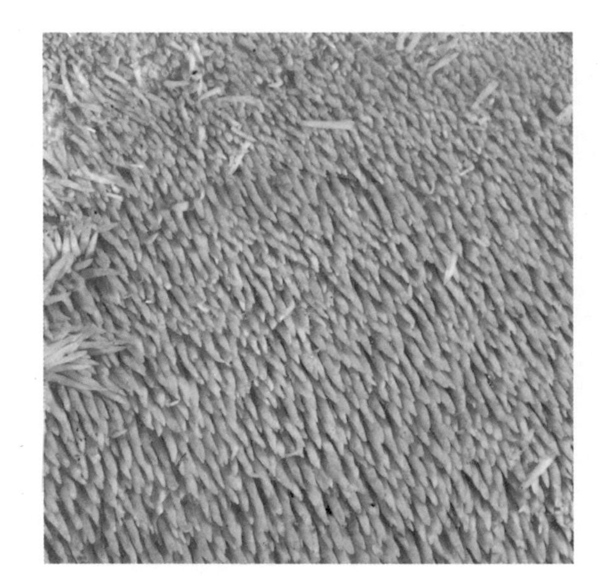

The Hedgehog Fungus

Hydnum repandum

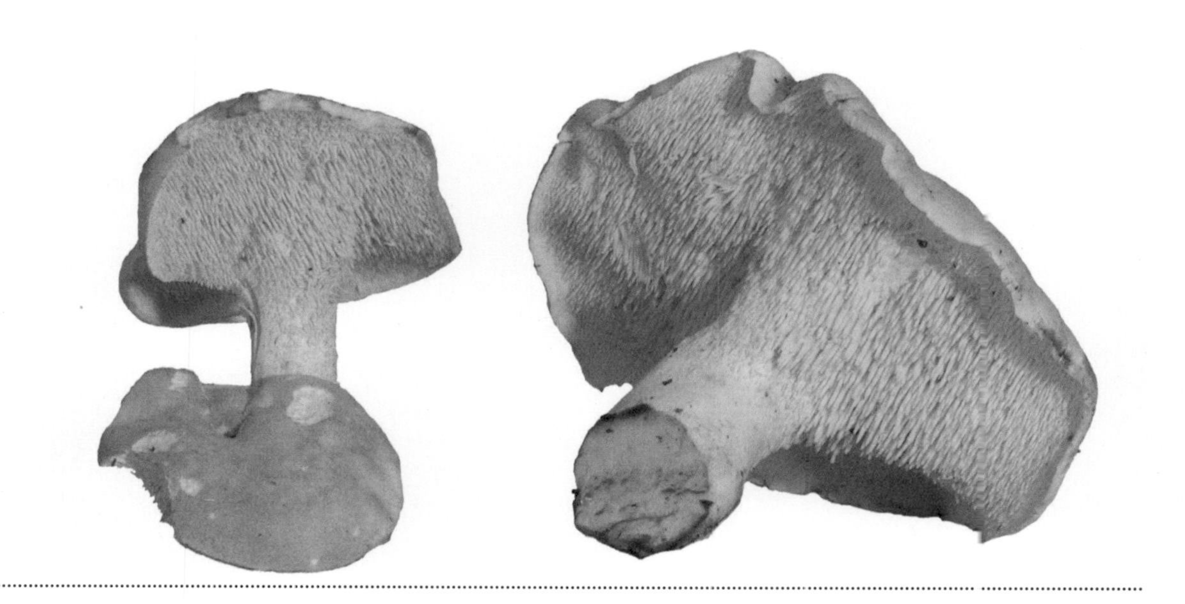

There are other mushrooms with spines but the only one of real culinary interest is the hedgehog fungus. The hedgehog fungus is a great delicacy. It grows on the ground but never on trees.

The spines are the key identification marks of the hedgehog fungus. The flesh is matte white and changes color in places to a yellow-brown or rusty yellow.

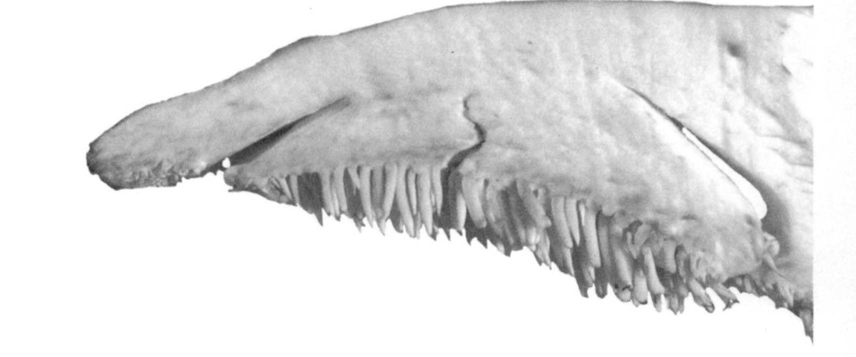

Some hedgehog fungi grow in a very irregular fashion, up to the size of two fists. Nevertheless, all hedgehog fungi have distinctive spines. The specimens here show the entire color range of the cap. No other mushroom with spines has these colors.

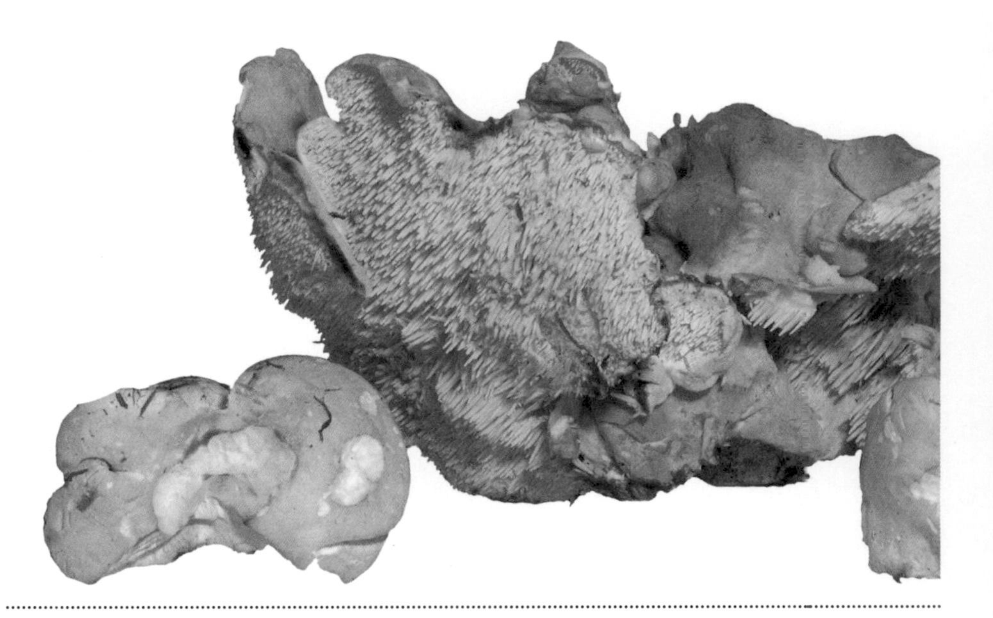

Take only specimens where the spines are clearly visible. The hedgehog fungi pictured here are the perfect size for the kitchen. The cap size is two and three-quarters to four inches (seven to ten centimeters) across.

Cap color bar

Positive ID Checklist

The Hedgehog Fungus

- ☑ Spines clearly visible
- ☑ Flesh matte white when freshly cut
- ☑ In places, flesh changes color to yellow-brown or rusty yellow
- ☑ Found on ground but not on trees
- ☑ Cap color matches cap color bar (above)

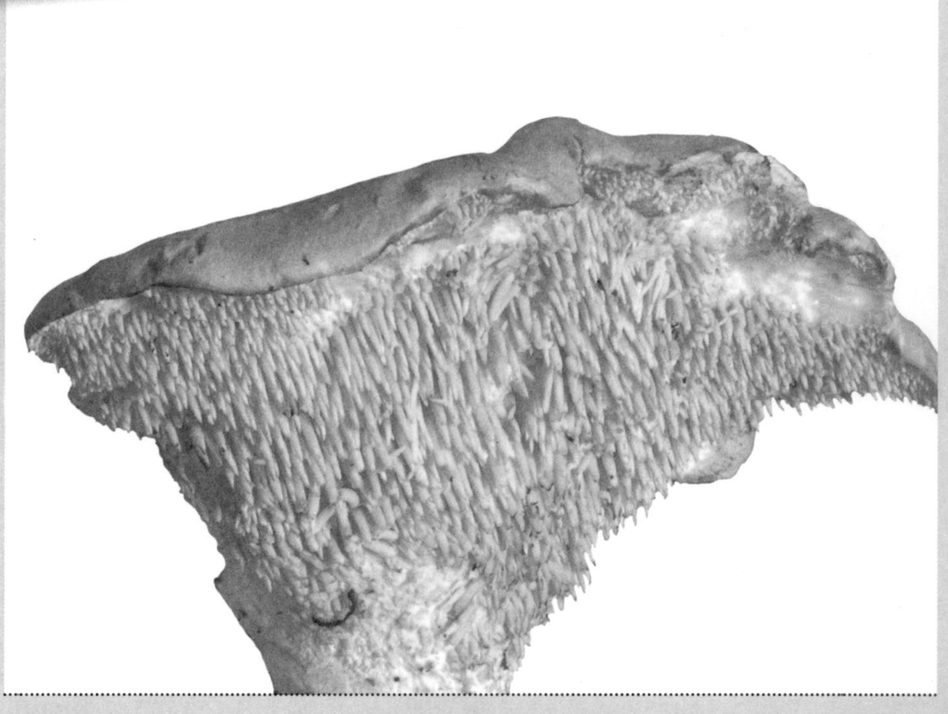

Avg. size across cap:	two and a half inches (six centimeters)
Season:	August to October
Habitat:	Habitat: Coniferous and deciduous woods
Tip:	Avoid older specimens as they can taste slightly bitter
Culinary rating:	10 out of 10

Repandum means "arching upward" (refers to the way the cap often reveals its underside)

Mavericks

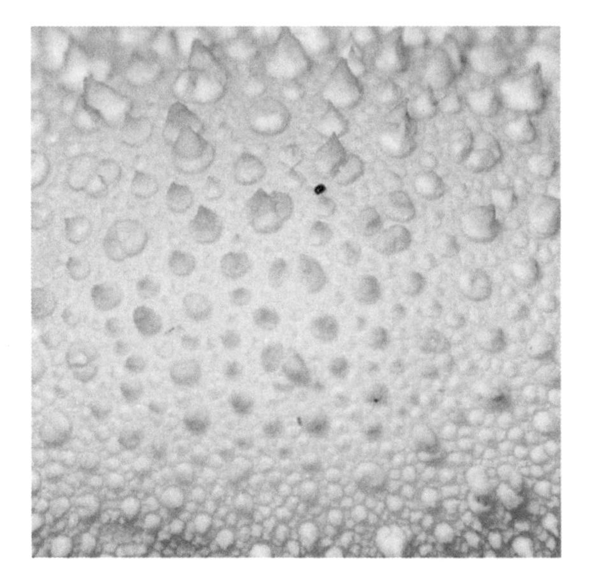

The Morel

Morchella esculenta

Morels open the mushroom season with fanfare. They might start fruiting in southern regions in January, and follow the spring north to fruit in May and early June in colder climates and higher altitudes. In certain regions of the continent, they are so prized that the details of sites where they are found are left in wills or passed on from the deathbed. They always appear on the same sites. Mycologists distinguish between various species and subspecies of morels. From the identification and culinary point of view, this doesn't matter because all morels share the same essential features and are equally delicious. Although you might stumble across morels in your garden, as a rule deciduous or coniferous woods and railway embankments, sandy soils, meadows, and ash trees are your best bet to start your quest for morels.

The morel is one of the hardest mushrooms to find, so study the landscape for clues. The scene above looks distinctly "morely" because there is a mix of old and young ash trees.

Middle-aged ash trees are easily identified by their diamond-patterned bark, which resembles the pitted and ridged structure of the morel.

Ash seeds and the characteristic black buds.

The bark of young ash trees is smoother, but the diamond pattern is discernible.

When these flowers appear, the morel season is in full swing.

Left: Coltsfoot
Right: White trillium.

Left: Marsh marigold
Right: Violet.

The cap is distinctly pitted and ridged. The pits and ridges must be clearly discernible, which is when the morel is about two inches (five centimeters) in height.

The form of the cap varies and so do the colors of cap and stem. The colors of the cap range from off-white (when young) to grayish, yellowish brown, brown, and blackish brown.

The average size of morels is about two to four inches (five to ten centimeters) in height, but they can grow up to eight to ten inches (twenty to twenty-five centimeters).

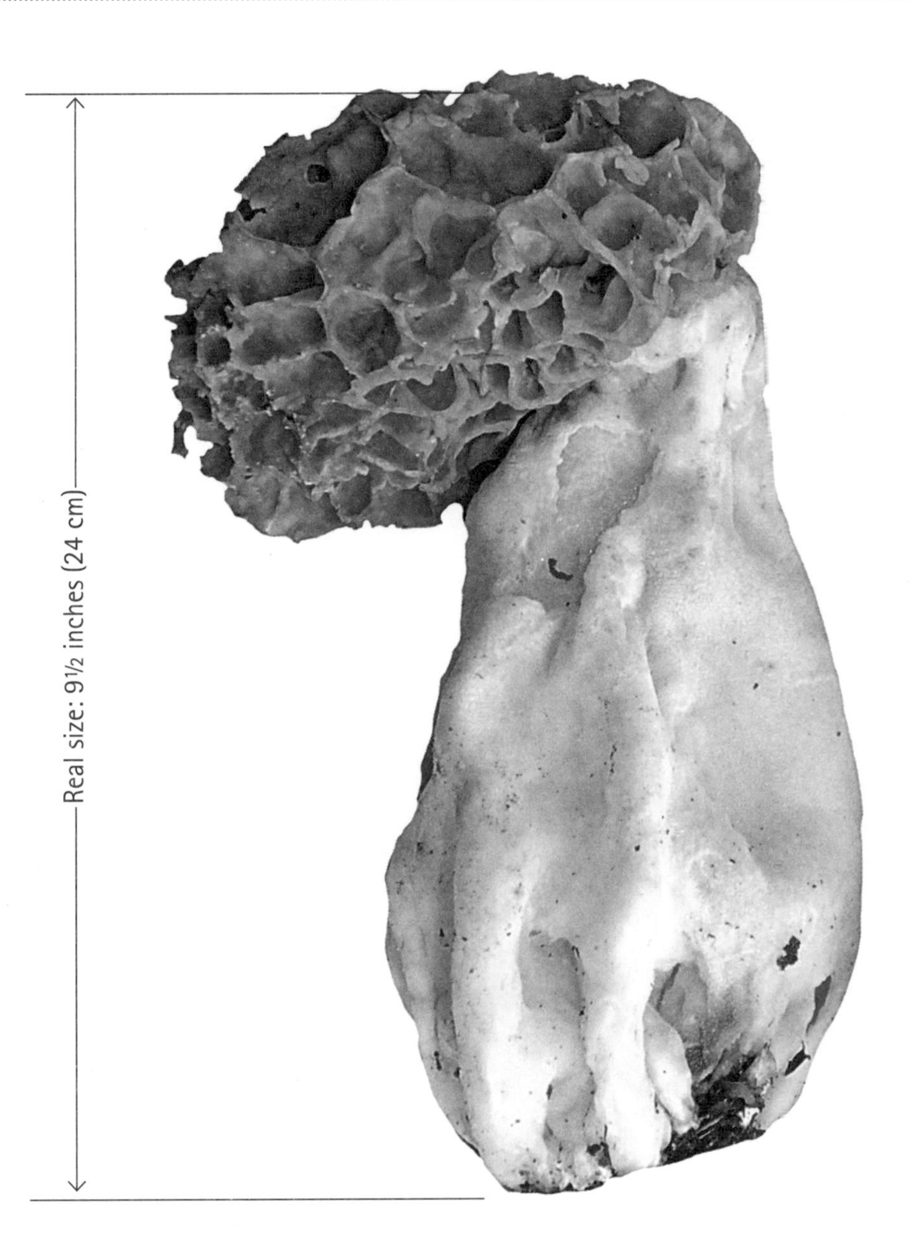

The mushroom on the left is too small to identify. It probably *is* a morel, but the pits and ridges can't be seen clearly; therefore you must give it time to grow. There is no doubt about the pits and ridges on the mushroom on the right.

Cap and stem are hollow.

All morels feature:

- Pits and ridges
- Hollow cap and stem
- Symmetry when cut in the middle

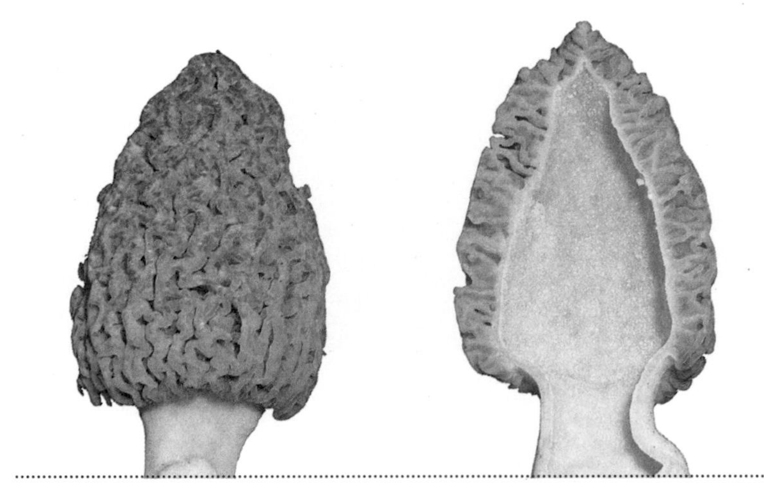

Pits and ridges: These are sometimes also referred to as "honeycomb pattern" or "polygonal cavities."

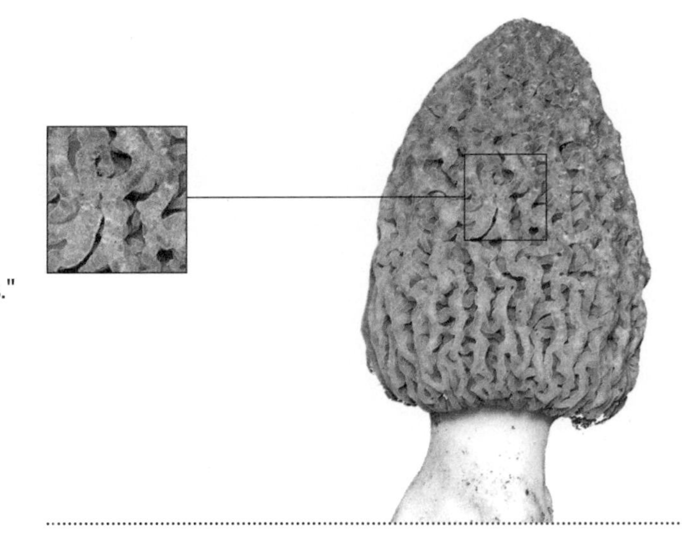

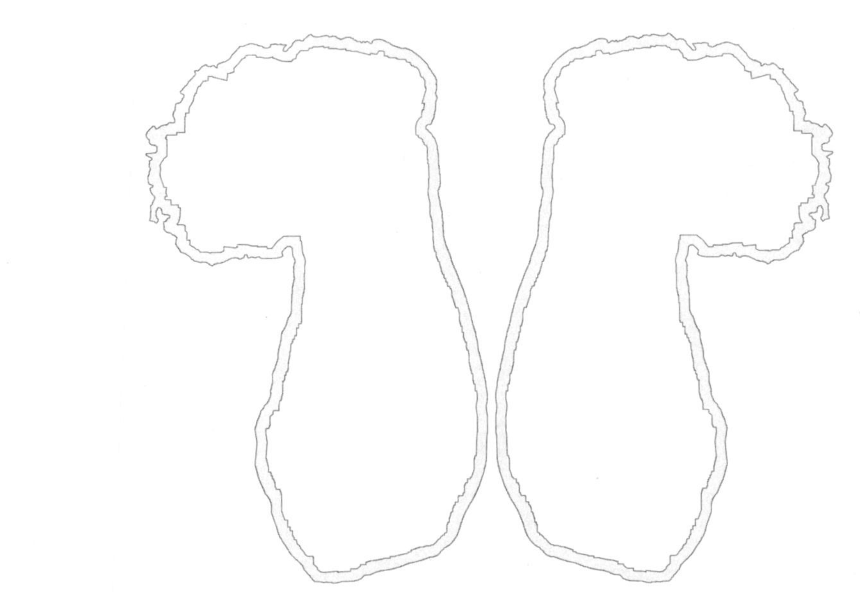

Morels are always symmetrical when cut in the middle.

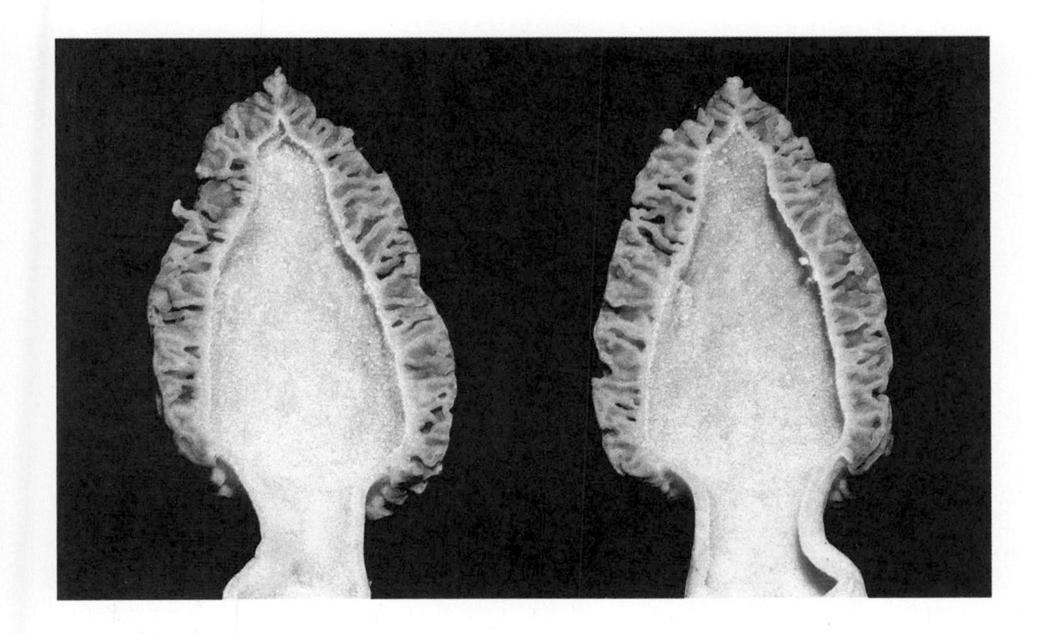

Cap color bar

Positive ID Checklist

The Morel

- ☑ Appears February to May
- ☑ Cap is pitted and ridged
- ☑ Completely hollow cap
- ☑ Completely hollow stem
- ☑ Symmetrical when cut through the middle
- ☑ Cap matches cap color bar (above)

Avg. size of cap height:	one to two inches (three to five centimeters)
Season:	January to June
Habitat:	Meadows, embankments, gardens
Tip:	Tastes best when dried
Culinary rating:	10 out of 10

Esculenta means "edible"

The Common Puffball

Lycoperdon perlatum

The common puffball is a curious little mushroom. One cluster of common puffballs might number merely four specimens; the next, about forty. The common puffball is a welcome addition to any mix of mushrooms. Cooked on its own, it is not everyone's cup of tea because of its distinctive taste. The giant puffball is another matter and as good a reason as you'll get for a dinner party.

The common puffball seems to feel at home everywhere and turns up in some unexpected places.

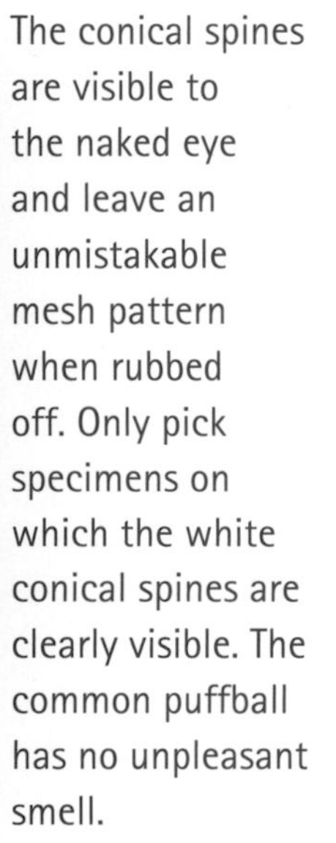

The conical spines are visible to the naked eye and leave an unmistakable mesh pattern when rubbed off. Only pick specimens on which the white conical spines are clearly visible. The common puffball has no unpleasant smell.

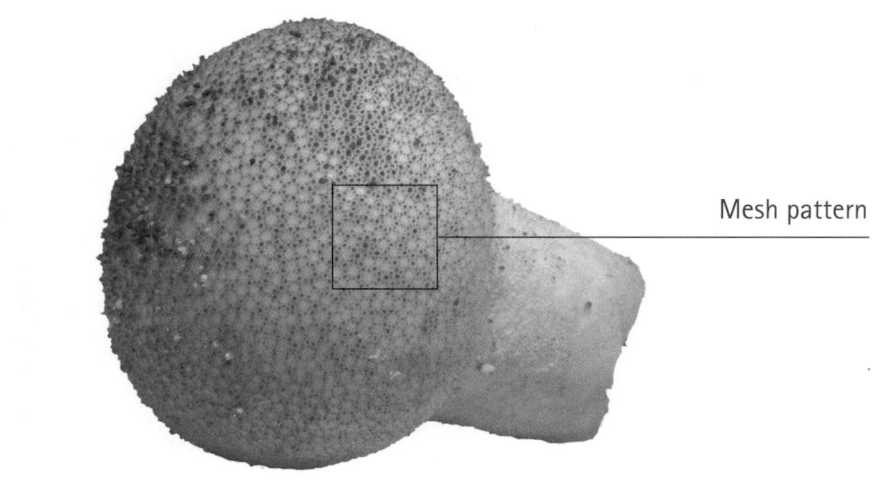

These common puffballs are the perfect size (one to two inches [three to five centimeters] high) and condition. When cut, the inside is all white and firm.

Only when the flesh is uniformly white and firm is the mushroom all right to use.

A cousin of the common puffball is the pestle puffball. It, too, has spines but these are finer than those of the common puffball. It is pestle-shaped and it grows larger (up to eight inches [twenty centimeters] high) than the common puffball. It has no smell. If the flesh is uniformly white and firm, it is good to eat.

The big brother of the common puffball is the giant puffball. "Giant" means it can grow to one yard (about a meter) in diameter.

From a distance, it looks all white but it can have a light yellow or yellow-brown tinge.

The surface is smooth and feels rather like suede. The flesh must be white and firm. Only then is the giant puffball good to eat.

Cap color bar

Positive ID Checklist

The Common Puffball

- Conical spines
- Flesh must be all white
- Flesh must be firm
- Mesh pattern when spines are rubbed off
- Color white to off-white
- Found in groups
- No unpleasant smell
- Cap color matches cap color bar (above)

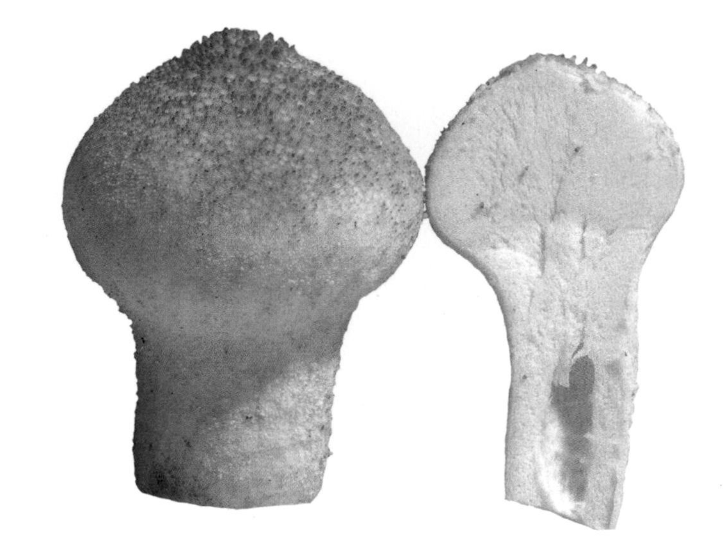

Avg. size across cap:	Size of a golf ball
Season:	July to October
Habitat:	Seems to feel at home everywhere
Tip:	Giant puffballs always grow on the same spot
Culinary rating:	9 out of 10

Perlatum means "widely spread"

The Hen of the Woods

Grifola frondosa

For thousands of years, the hen of the woods mushroom has been prized for its medicinal and culinary value in China and Japan. "Maitake," the Japanese name for the hen of the woods, means "dancing mushroom," perhaps because those who found a hen of the woods started to dance with joy at discovering such a highly prized mushroom. The hen of the woods (not to be confused with the chicken of the woods) is a mushroom that can be cultivated commercially. However, the wild hen of the woods mushroom you pick is infinitely superior because, firstly, you found it yourself; and secondly it is as different from the farmed version as is a wild salmon from a farmed one. Finding a hen of the woods is indeed a reason to dance: a culinary event of the first order. And it's good for you, too. There aren't many things in life of which that can be said!

The hen of the woods is a cluster of fan-shaped overlapping caps.

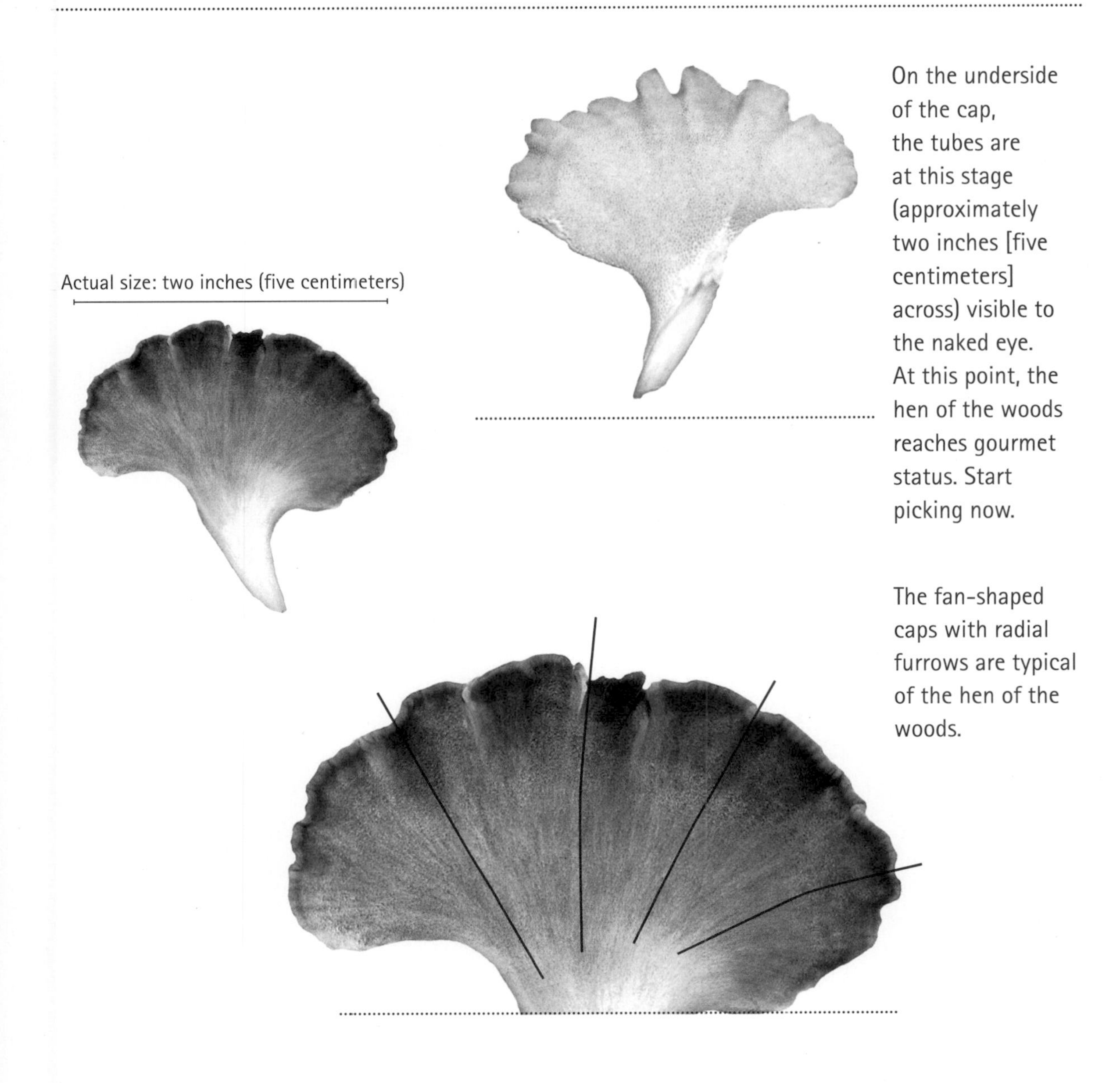

On the underside of the cap, the tubes are at this stage (approximately two inches [five centimeters] across) visible to the naked eye. At this point, the hen of the woods reaches gourmet status. Start picking now.

The fan-shaped caps with radial furrows are typical of the hen of the woods.

The tubes are clearly visible at this stage.

The cross-section shows a cauliflower-like structure with one central stem. The widest part of this particular specimen is approx. twelve inches (thirty centimeters).

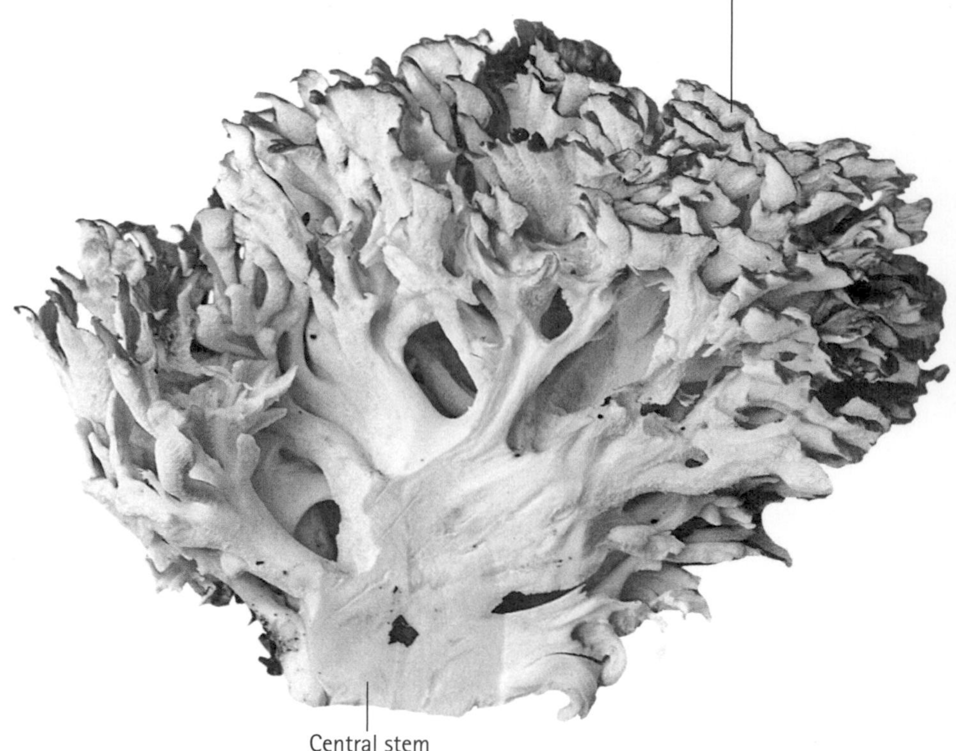

On older specimens, the tubes are larger in diameter but remain whitish. As the tubes grow larger, they almost look like spines—but tubes they are.

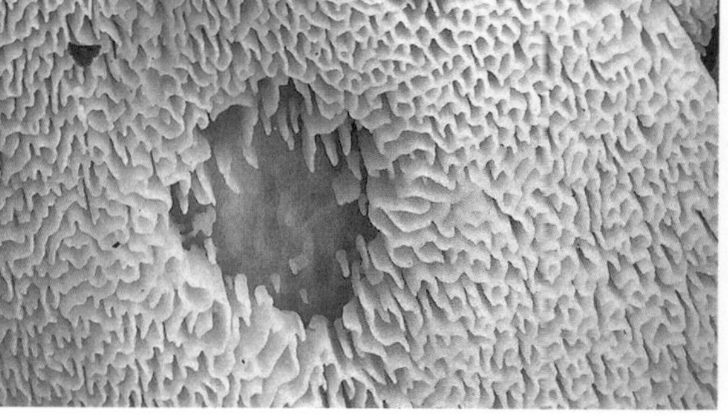

The main colors of the hen of the woods range from off-white, gray-beige, gray-brown, gray-black, brown-black, to light brown-black. The hen of the woods can grow up to thirty-two inches (eighty centimeters) across.

The culinary value of the hen of the woods depends on its age. Some young specimens can grow very big very fast. These are young specimens: white, firm, fibrous flesh with a pleasant smell. Size is not an indicator of age. Older specimens have a distinct unpleasant odor. Their caps get more and more flabby and the edges crumble.

Cap color bar

Positive ID Checklist

The Hen of the Woods

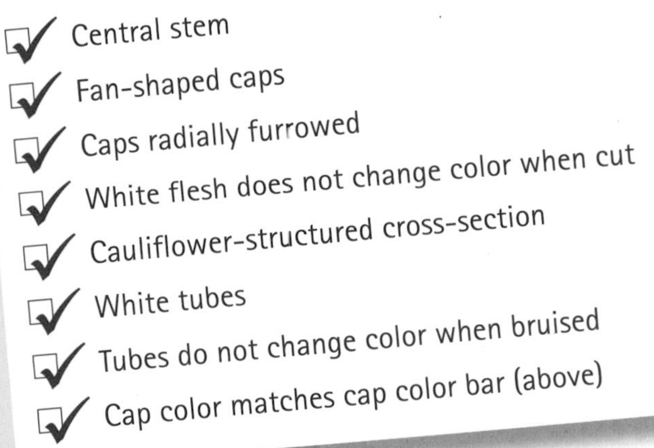

Avg. size:	twelve inches (thirty centimeters) across but can grow up to thirty-two inches (eighty centimeters)
Season:	September to October
Habitat:	Mainly at the base of oaks but also on other deciduous trees
Tip:	Grows on the same spot for many years
Culinary rating:	10 out of 10

Grifola frondosa means "leafy griffin"

The Horn of Plenty

Craterellus cornucopioides

The horn of plenty is a real challenge to the mushroom forager. Growing low on the forest floor, these mushrooms are so well camouflaged that you can look at masses of them and not see any at all. If you spot one, there will be more: Proceed with the utmost care because they are easily trampled. The horn of plenty is without a shadow of a doubt one of the most delicious mushrooms in the world. But it has its price. Cleaning the horn of plenty can be quite a job but the reward is well worth the effort. Besides, when everybody joins the cleaning party, it's fun.

The horn of plenty is a funnel, the outer side of which looks smooth but is in fact slightly wrinkled. The color on the outer side is ash-gray, gray, or pale gray with a bluish or lilac tint.

The inside of the funnel is brown, brown-gray, soot gray, or black. The surface is scurfy or flaky and the top edge is curled under. In older specimens, the top is wavy and split.

The horn of plenty from above in dry conditions.

The horn of plenty from above in wet conditions.

The funnel shape and the curled-under top are typical.

This seems to be a rather battered group of horn of plenty. Some of the tops have already split. Don't be put off—they still taste good.

Outer funnel

Positive ID Checklist

The Horn of Plenty

- Found in groups
- Funnel-shaped
- Smooth-looking but slightly wrinkled outer funnel
- Scurfy inner funnel
- Outer funnel matches color range bar (above)
- Inner funnel matches color bar (below)

Avg. size across cap:	three quarters of an inch to an inch (two to two and a half centimeters)
Season:	August to November
Habitat:	Near oak and beech
Tip:	When cleaning, always split down the middle
Culinary rating:	10 out of 10

Cornucopioides refers to a "cornucopia," i.e., horn of plenty

Inner funnel

The Cauliflower Mushroom

Sparassis crispa

This is the mushroom of superlatives. For many people it is the very best of all. It may be a bit chewy but each bite releases a unique culinary sensation. If you fall for it, you'll be a dedicated cauliflower mushroom hunter forever. It resembles a cauliflower or a sponge. Once you learn to recognize its spicy smell, you could identify the cauliflower mushroom blindfolded. And then there is the size: It can grow so big that people simply overlook it. This specimen weighed in at 22.5 lb (10.2 kg).

Not all specimens grow to giant sizes. This one here is about the size of two fists. Its cauliflower- or sponge-like appearance and the curved lobes are the key identification marks. There is nothing pointed or jagged in the cauliflower mushroom. The color ranges from creamy white to light brown. If it turns any browner than the specimen here, it has gone off.

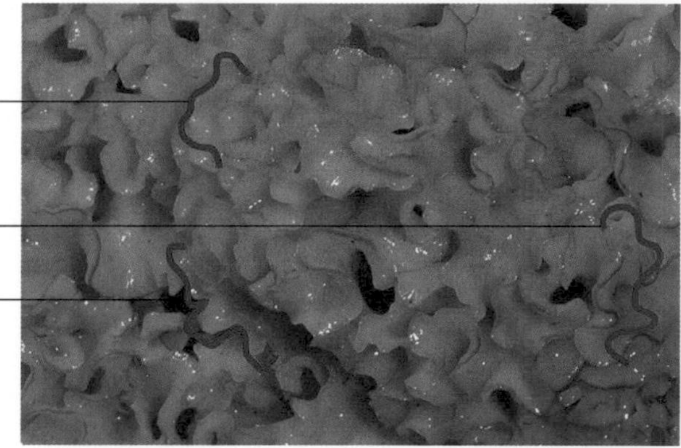

Color range bar

Positive ID Checklist

The Cauliflower Mushroom

- ☑ No gills, pores, tubes, or spines
- ☑ No stem
- ☑ Looks like a cauliflower or sponge
- ☑ Lobes
- ☑ Color matches color bar (above)

Avg. size:	twelve inches (thirty centimeters) across (but can grow up to a meter)
Season:	September to October
Habitat:	Mainly near pine or pine stumps
Tip:	Grows on the same spot for many years
Culinary rating:	10 out of 10

Crispa means "frizzy"

Trees and Mushrooms

Beech, oak, birch, pine, larch, and ash are easily identifiable and are excellent pointers to certain mushrooms.

Beech
Especially:
king bolete, horn of plenty.

Oak
Especially:
king bolete.

Birch
Especially:
birch bolete, orange birch bolete.

Pine
Especially: cauliflower mushroom.

Larch
Especially: larch bolete.

Ash
Especially: morel.

Handling, Storage, and Cooking

Cleaning

The first rough cleaning of the mushroom should be done in the woods. Once you're at home, you must verify your mushrooms against the positive ID checklists and then—and only then—can you start the fine-cleaning.

Do not wet mushrooms to clean them. Brush them or wipe them with a cloth. Part of the cleaning process is to check the quality. Evict residents, cut away soft and soggy tubes, and cut the remaining mushroom into bite-size pieces.

Basic preparation for fresh or frozen mushrooms

There's a lot of leeway in these basic preparations. The amounts stated can vary considerably and you'll still get an excellent result. And the beauty of it is that you can't overcook mushrooms. The key to culinary success is the quality of the mushrooms.

Serves 4

What you need

Essential:

10.5–21 oz (300–600 g) fresh or frozen mushrooms
3 1/3–4 1/5 cups (800 ml–1 liter) water
2/5–4/5 cups (100–200 ml) dry white wine
3/4 vegetable stock cube
1 shallot or small onion, chopped
2/5–4/5 cups (100–200 ml) cream
0.3–0.7 oz (10–20 g) chives or parsley

Optional:

A mix of finely chopped vegetables, e.g., carrots, zucchini, and celery: approx. 2.4 oz (70 g)
0.3–0.7 oz (10–20 g) chervil

What you do

1. Put water, white wine, vegetable stock, and shallot into a pan.
2. Do not thaw frozen mushrooms. Add frozen or fresh mushrooms and bring to a boil.
3. Reduce heat and simmer until approximately 2/3 of liquid has evaporated.
4. Add cream.
5. Simmer again until approximately half the liquid has evaporated. Stir occasionally. If you feel there's too much liquid, simmer until it reduces still further. The less liquid, the more intense the taste, and vice versa. Add water or cream to taste.
6. Add salt to taste.

   You now have a wonderful mushroom dish that you can serve with meat (chops, steak, etc.) or on its own with pasta or rice. Sprinkle with chives, parsley, or chervil and the finely chopped vegetables.

Basic preparation of dried mushrooms

Serves 4

What you need

Essential:

1–2 oz (30–60 g) dried mushrooms
1 quart (1 liter) lukewarm water
1 shallot or small onion, chopped
4/5 cups (200 ml) dry white wine
3/4 vegetable stock cube
2/3–4/5 cups (150–200 ml) cream
0.3–0.7 oz (10–20 g) chives or parsley

Optional:

A mix of finely chopped vegetables, e.g., carrots, zucchini, and celery: approx. 2.4 oz (70 g)
0.3–0.7 oz (10–20 g) chervil

What you do

1. Take the dried mushrooms and put in a bowl or pitcher. Add 1 quart of lukewarm water. Soak for 1½ hours (until the mushrooms float in brown liquid).
2. Put wine, shallot, vegetable stock, mushrooms, and the brown liquid into a pan.
3. Bring to a boil for 10 seconds. Reduce heat and simmer until 2/3 of the liquid has evaporated.
4. Add the cream.
5. Simmer until approximately half the liquid has evaporated. Stir occasionally. If there is too much liquid, simmer until further reduced. The less liquid, the more intense the flavor, and vice versa.
6. Add water or cream to taste.
7. Add salt to taste.

You now have a wonderful mushroom dish that you can serve with meat (chops, steak, etc.) or on its own with pasta or rice. Sprinkle with chives, parsley, or chervil and the finely chopped vegetables.

For more about these basic preparations and on cooking mushrooms, visit the website www.mushrooming.co.uk.

Germs and special cases

Kills all known germs

There is a very low risk of catching something nasty from a mushroom. To reduce that risk to zero, do not eat wild mushrooms raw and in the cooking process, increase the heat at one stage so that the dish either boils for a couple of seconds or sizzles in the butter. For that brief moment, turn the mushrooms in the pan so that they're exposed to the heat on all sides.

Chanterelle

The chanterelle must not be dried because it goes chewy. It must not be frozen without prior cooking because it will turn bitter.

Horn of Plenty

Some people like it on its own. Others use it only as a spice to add to a mix of mushrooms. In order to find out what you like best, dry and freeze separately.

Cauliflower mushroom

The cauliflower mushroom is also very distinct in taste and perhaps best on its own. Dry or freeze separately.

More on the website: www.mushrooming.co.uk

Storage

Drying

Drying is the classic way to store mushrooms. This method actually intensifies the taste of the mushrooms and they'll endure for years if you follow the procedure correctly. Do not dry anything you wouldn't eat fresh. Drying cannot improve the overall quality of your mushrooms!

There are various methods of drying mushrooms, but there is only one way that guarantees consistently high-quality results and that is the only one I recommend: a dehydrator. Whatever the make, the principle is the same: warm air circulation. The sliced mushrooms are placed on trays and then dried in the dehydrator.

Dried mushrooms are best stored in any airtight container or a sealed plastic bag and then kept in a dark place. Mix all the species except the cauliflower mushroom. The more species in the mix, the better the taste. The cauliflower mushroom should be dried and stored separately. **Do not dry chanterelles.** All other mushrooms in this book are suitable for drying. Dried mushrooms should be crackle-dry and snap when broken.

More on the website:
www.mushrooming.co.uk

Freezing

Freezing is the other method. I recommend only freezing vacuum-packed mushrooms. Frozen mushrooms retain their color, texture, and taste.

Never thaw frozen mushrooms!
Put frozen mushrooms in sizzling butter or hot water to retain their texture.

- Mix all the species except the cauliflower mushroom. Again: The more species in a mix, the better.
- I recommend freezing some king bolete and horn of plenty, separately. You might want to add just a few of them for a particular dish or use the king bolete for a tasty starter.

The cauliflower mushroom should always be frozen separately.

The chanterelle **must be cooked** before freezing. All other mushrooms in this book can be frozen raw.

More on the website:
www.mushrooming.co.uk

Drying or freezing?
Either way, use **only perfectly fresh mushrooms.** Drying mushrooms gives them that deep, strong mushroom taste. The frozen mushroom is more subtle, and the colors of frozen mushrooms are as bright as on the day they were picked, which adds considerably to the pleasure of eating. Dried and frozen mushrooms can be mixed to get the best of both worlds.

More on the website:
www.mushrooming.co.uk

Index

The 25 best-to-eat and most common mushrooms

Field Mushroom

Wood Blewit

Shaggy Mane

Parasol

Shaggy Parasol

Oyster Mushroom

Charcoal Burner

Amethyst Deceiver

King Bolete

Red Cracked Bolete

Larch Bolete

Slippery Jack

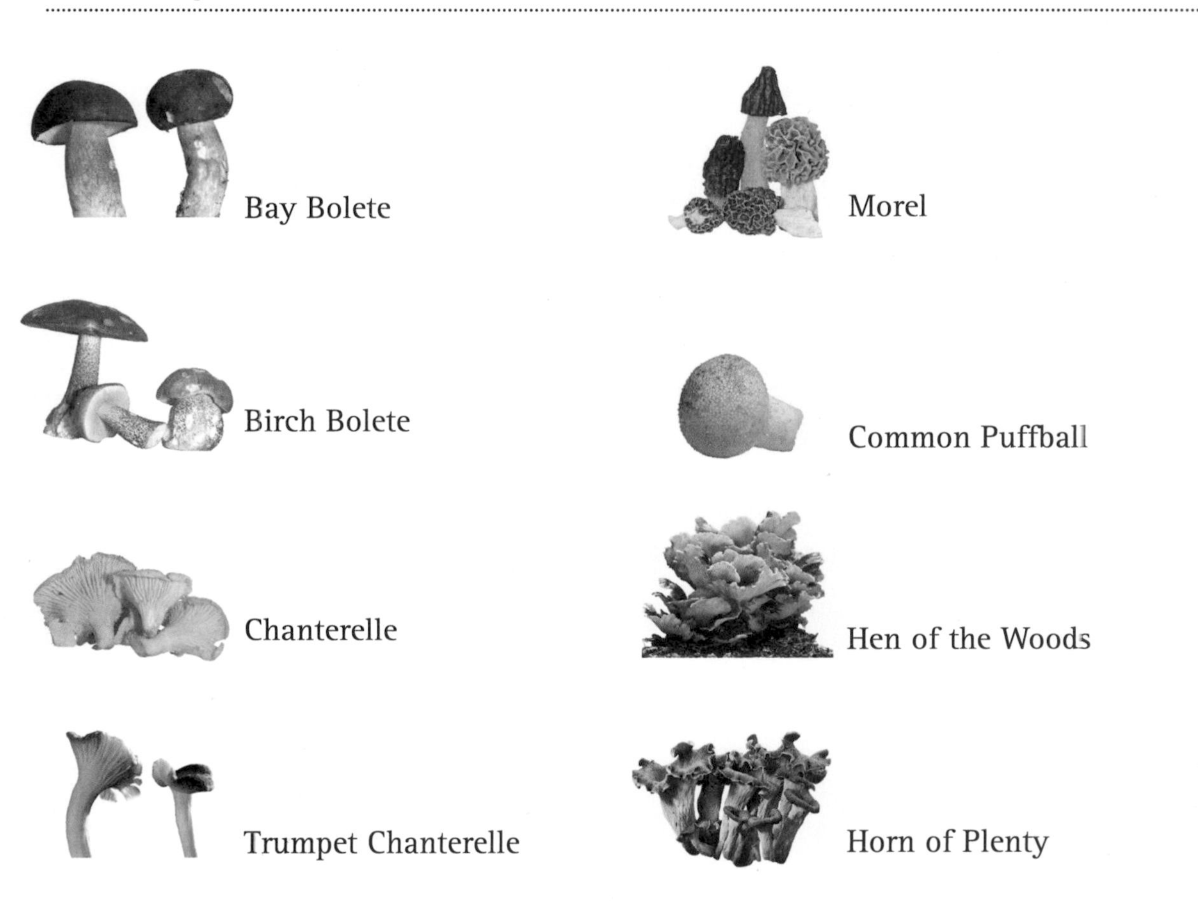

Bay Bolete

Birch Bolete

Chanterelle

Trumpet Chanterelle

Morel

Common Puffball

Hen of the Woods

Horn of Plenty

Hedgehog Fungus

Cauliflower Mushroom